EARTHQUAKE SPECTRA AND DESIGN

Monograph Series

Engineering Monographs on Earthquake Criteria, Structural Design, and Strong Motion Records

Coordinating Editor, Mihran S. Agbabian

Monographs Available

Reading and Interpreting Strong Motion Accelerograms, by Donald E. Hudson

Dynamics of Structures—A Primer, by Anil K. Chopra

Earthquake Spectra and Design, by Nathan M. Newmark and William J. Hall

EARTHQUAKE SPECTRA

AND DESIGN

by **N. M. Newmark**
and **W. J. Hall**
Department of Civil Engineering
University of Illinois at Urbana-Champaign

EARTHQUAKE ENGINEERING RESEARCH INSTITUTE

Published by

The Earthquake Engineering Research Institute, whose objectives are the advancement of the science and practice of earthquake engineering and the solution of national earthquake engineering problems.

This is volume three of a series titled: Engineering Monographs on Earthquake Criteria, Structural Design, and Strong Motion Records.

The publication of this monograph was supported by a grant from the *National Science Foundation*.

Library of Congress Catalog Card Number 82-71183
ISBN 0-943198-22-4

This monograph may be obtained from:
   Earthquake Engineering Research Institute
   2620 Telegraph Avenue
   Berkeley, California 94704

The views expressed in this monograph are those of the authors and do not necessarily represent the views or policies of the Earthquake Engineering Research Institute or of the National Science Foundation.

TRIBUTE TO NATHAN M. NEWMARK

The final stage of preparation of this monograph was interrupted by the untimely death of one of the authors, Nathan M. Newmark (1910-1981). A long-time member of the Earthquake Engineering Research Institute, Professor Newmark was one of the pioneers in the field of earthquake engineering. With his students at the University of Illinois, he carried on research on the effects of earthquake ground shaking on structures and on the design of structures to resist seismic stresses and strains. He was also very active in engineering consultation on aseismic design of major projects including: nuclear power plants throughout the United States and in foreign countries; the high-rise Latino Americana Tower building in Mexico City; the San Francisco Bay Area Rapid Transit System; the Alaska Oil Pipe Line; and others. His advice was frequently sought by governmental agencies on engineering problems of national importance. Many of his former students are themselves now prominent in earthquake engineering research and practice, so his influence will continue to be felt in civil engineering. During his career Dr. Newmark received many honors and awards, including the National Medal of Science—through his death the engineering profession has lost one of its most eminent members.

PAUL C. JENNINGS
President, EERI

Latino-American Tower. When constructed in the early 1950s, this 43-story steel-frame skyscraper, rising 456 ft. above street level with a superimposed 138 ft. television tower, was the tallest building south of the U.S. border. The principal designers were A. Zeevaert and L. Zeevaert, with the seismic design, based on the principles of modern dynamic analysis, being the responsibility of N. M. Newmark. In 1957 the building withstood without damage the largest earthquake on record in Mexico (MM VII–VIII); displacement measurements at three floor levels documented that the building responded as designed.

FOREWORD

The occurrence of earthquakes poses a hazard to cities that can lead to disaster unless appropriate engineering countermeasures are employed. Recent earthquake disasters with high death tolls: in Guatemala, 1976 (20,000); Tangshan, China, 1976 (500,000); Iran, 1978 (19,000); Algiers, 1980 (10,000); Italy, 1980 (3500) demonstrate the great advantages that could be gained by earthquake resistant construction. To provide an adequate degree of safety at an affordable cost requires a high level of expertise in earthquake engineering and this in turn requires an extensive knowledge of the properties of strong earthquakes and of the dynamics of structures that are moved by ground shaking. To achieve this it is necessary for relevant information to be published in an appropriate form.

This monograph by N. M. Newmark and W. J. Hall on earthquake resistant design considerations is the third in a projected series of monographs on different aspects of earthquake engineering. The monographs are by experts especially qualified to prepare expositions of the subjects. Each monograph covers a single topic, with more thorough treatment than would be given to it in a textbook on earthquake engineering. The monograph series grew out of the seminars on earthquake engineering that were organized by the Earthquake Engineering Research Institute and presented to some 2,000 engineers. The seminars were given in 8 localities which had requested them: Los Angeles, San Francisco, Chicago, Washington, D.C., Seattle, St. Louis, Puerto Rico, and Houston. The seminars were aimed at acquainting engineers, building officials and members of government agencies with the basics of earthquake engineering. In the course of these seminars it became apparent that a more detailed written presentation would be of value to those wishing to study earthquake engineering, and this led to the monograph project. The present monograph discusses important aspects of structural design that are different in seismic engineering than in designing to resist gravity loads.

The EERI monograph project, and also the seminar series, were supported by the National Science Foundation. EERI member M. S. Agbabian served as Coordinator of the seminar series and also is serving as Coordinator of the monograph project. Technical editor for the series is J. W. Athey. Each monograph is reviewed by the members of the Monograph Committee: M. S. Agbabian, G. V. Berg, R. W. Clough, H. J. Degenkolb, G. W. Housner, and C. W. Pinkham, with the objective of maintaining a high standard of presentation.

> GEORGE W. HOUSNER
> *Chairman, Monograph Committee*

Pasadena, California
March, 1982

PREFACE

Recent major earthquakes in Alaska (1964), San Fernando, California (1971), and Peru (1970), with their accompanying massive land and submarine slides, attest to the need for considering such natural hazards, their possibility of occurrence and their consequences. Because our expanding population is concentrated in large metropolitan centers with a proliferation of man-made structures and facilities, the number of incidents and extent of the consequences (loss of life, injury, and loss of property or damage) from such disasters can be expected to increase in the years ahead. Even in geographical areas where seismic risk is assumed to be low, as in the eastern United States, consequences of a possible large earthquake are serious and require careful consideration.

An even greater consequence is that the technology of our society requires the use of structures and facilities whose damage or destruction by natural hazards could be very serious, for example nuclear power plants, large dams, and certain pipelines, lifelines, and industrial facilities. Damage to such "critical facilities"—which include hospitals, emergency service facilities and essential utilities—can affect the public well-being through loss of life, large financial loss, or degradation of the environment if they were to fail functionally. Some of these facilities must be designed to remain operable immediately after an incident to provide life-support services to the communities affected.

The goal of earthquake engineering is to ensure that in the event of an earthquake there will be no serious injury or loss of life. From the physical standpoint, the general purpose of earthquake-resistant design is to provide a structure capable of resisting ground motions expected to occur during the lifetime of the structure. In this case the objective centers partly on economics, as well as life safety, in that the design is made in such a manner that ideally the cost of repair of earthquake damage will not exceed the increased design, construction, and financing costs necessary to have prevented the damage in the first place. In the case of industrial facilities, the goal also is that of minimizing or eliminating operational disruptions. Moreover, through attention to design and construction, another objective is that of mitigating serious failure or collapse in the event of a major earthquake, i.e., larger than the seismic hazard for which the design was made; this is to be done however rare the probability of its occurrence.

For building-type structures, seismic design procedures, such as those included in a building code, usually prevail and are enforceable under the applicable jurisdictional authority. The seismic provisions of standard building codes generally center around the philosophy expressed in the preceding paragraph. For critical facilities, special seismic design criteria are developed as a part of the design process. In such cases, comprehensive geological and seismological investigations are usually required. The development of seismic design criteria that are sound in principle for such situations requires close cooperation between the geologist, seismologist, earthquake engineer, architect and client throughout the design process if a viable and economically satisfactory project is to be achieved.

The purpose of this monograph is to describe briefly some of the concepts and procedures underlying modern earthquake engineering, especially as it applies to building structures. The presentation was developed so as to convey in some logical sequence the material presented by the authors in the EERI lecture series; in certain areas brief updating has been added. In order to provide a self-contained monograph, the introductory material presented under General Concepts and Earthquake Ground Motions leads logically into the discussion of Design Response Spectra, one of the basic concepts underlying modern earthquake engineering analysis. The section on Dynamic Structual Analysis Procedures contains a description of the basic modal analysis and the equivalent lateral force procedures that are normally employed as a part of the design process. Thereafter follows a brief description of topics that require special design consideration and a brief introduction to the recently developed Applied Technology Council Provisions. The seismic design procedures discussed herein are restricted essentially to buildings, although some aspects of the topics discussed are applicable to facilities generally.

TABLE OF CONTENTS

	PAGE
General Concepts	13
Introduction	13
Seismic Design and Analysis Concepts	14
Basic Function of Design Codes	15
Earthquake Ground Motions	20
General Observations	20
Site Amplification and Modification by Presence of Structure	22
Actual Versus Effective Earthquake Motions	24
Design Response Spectra	29
Spectrum Concepts	29
Modification of Spectra for Large Periods or Low Frequencies	45
Dynamic Structural Analysis Procedures	49
Introduction	49
Building Properties and Allowable Ductility Factor for Analysis	52
Mass and stiffness	52
Damping and ductility	53
Modal Analysis Procedure	57
Structural idealization	59
Modal periods and shapes	59
Modal responses	60
Total responses	62
Application to inelastic systems	63
Combined earthquake design responses	64
Equivalent Lateral Force Procedure	64
Fundamental period of vibration	65
Lateral forces	67
Story forces	69
Deflections and drifts	70
Earthquake design responses	71

Special Design Considerations........................ 73
   Torsion... 73
   Distribution of Shears............................... 74
   Base or Overturning Moments....................... 75
   Vertical Component of Ground Motion 76
   Combined Effects of Horizontal and Vertical Motions ... 76
   Effects of Gravity Loads 77
   Limitation and Choice of Lateral Force and
      Modal Analysis Methods 79

Applied Technology Council Provisions 84
   Background 84
   Basic Concepts of ATC-3 Provisions................. 86

Concluding Statement 97

Acknowledgment 98

References 99

Earthquake Spectra and Design

by
N. M. Newmark *and* **W. J. Hall**
University of Illinois at Urbana-Champaign

GENERAL CONCEPTS

Introduction

When a structure or a piece of equipment or instrumentation is subjected to earthquake motions, its base or support tends to move with the ground on which it is supported or with the element on which it rests. Since this motion is relatively rapid, it causes stresses and deformations in the item. If this component is rigid, it moves with the motion of its base, and the dynamic forces acting on it are nearly equal to those associated with the base accelerations. However, if the component is quite flexible, large relative motions or strains can be induced in the component because of the differential motions between the masses of the component and its base. In order to survive the dynamic motions, the element must be sufficiently strong as well as sufficiently ductile to resist the forces and deformations. In assessing seismic effects it should be remembered that the effects arising from seismic actions lead to changes in already existing effects, such as forces, moments, stresses, and strains arising from dead load, live load and thermal effects.

Unfortunately, the earthquake hazard for which an element or component should be designed is subject to a high degree of uncertainty. In only a few areas of the world are there relatively long periods of observations of strong earthquake motions. The effects on a structure, component, or element, depend not only on the earthquake motion to which it is subjected, but also on the properties of the element itself. Among the more important properties are the ability of the element to absorb energy within it or at interfaces between the element and its support, either due to damping or inelastic behavior, the period of vibration, and the strength or resistance.

This monograph begins with a discussion of the general concepts underlying earthquake-resistant design, including a brief description of the basic function of design codes. Thereafter follows a brief discussion of ground motions including some detailed discussion of the concept of effective acceleration or effective motion, i.e., the motion that controls in the design process. This section leads naturally into the next section dealing with basic concepts associated with design response spectra. At this point, earthquake-resistant response and analysis procedures are introduced with particular attention to basic modal analysis and equivalent lateral force procedures. Thereafter follows a discussion of selected topics requiring special design consideration. The monograph ends with an overview of the Applied Technology Council provisions that in all likelihood will form the basis of earthquake-resistant design provisions in building codes in the years ahead.

Seismic Design and Analysis Concepts

Once a structure has been laid out in plan and the size and strength of its various elements selected, then the analysis of the structure for given conditions of dynamic loading and foundation motion can be made by relatively well understood methods, even though the analysis may be a tedious and lengthy one for a complex system. However, unless the designer employs the so-called direct design procedure, such as that found in most current building codes, he is faced with the basic problem of the preliminary selection of the structural layout and element strength before he has a structure that can be analyzed.

In broad perspective, the steps which the designer normally must take in the case of earthquake-resistant structural design are as follows:

1. Select the design earthquake hazard with appropriate consideration of the acceptable risk.

2. Select the level of conservatism desired in the design. If nonlinear behavior is to be permitted, one may need to specify the allowable limits of deformation, or in some special cases the allowable probability of damage or failure. The selection of these items may depend in part on Step 3.

3. Select the type of layout of the structure and estimate its dynamic and static parameters. These parameters include (a) dynamic and static resistance, (b) natural frequencies or periods of vibration, (c) damping or energy absorption appropriate to the desired response, and (d) deformation that can be accommodated before failure or loss of function. These parameters may be assigned in a direct design procedure, or are subject to successive revision in more traditional procedures.

4. Through use of analysis, verify the adequacy of the structure selected and make any necessary changes in layout or element strength. Check to be sure that a compatible design exists for resisting all anticipated loadings, such as those arising from gravity, wind and earthquake. Steps 3 and 4 should be repeated until a satisfactory design is achieved.

5. If warranted, make a more accurate analysis of the final design and make further changes or refinements as may be necessary. Repeat Steps 2 to 4 if required.

In some cases an upper bound direct procedure may be used involving essentially only Steps 1, 2 and 3; most so-called pseudo static earthquake design procedures in building codes are intended to be of this type. In the case of reanalysis of an existing structure the process is essentially the same, except that the dimensions and properties are known or estimated, and iteration would be required primarily only if retrofit structural changes are to be made, i.e., if possible or economically feasible.

Basic Function of Design Codes

The designer, as well as others who have a responsibility for the final structure, must have some general method of knowing that gross errors have been avoided, as well as a basis of comparison to ensure that the design is adequate in an overall sense. It is the purpose of building codes and specifications to fill this need in part. However, it is not yet established that building codes can accomplish this task without introducing constraints and controls that may be a severe handicap in the development of new design concepts and procedures. Building codes embody the result of experience and judgment and therefore must deal at least implicitly, if not explicitly, with particular structural types and configurations. This discussion would not be complete without noting that the real purpose of building codes is to promulgate provisions that are intended to protect life—and to a lesser degree property—to the extent practical or to some perceived level of safety. Generally such provisions are minimum requirements for protection of the public, i.e., these are necessary but, depending on the circumstances, may not necessarily be sufficient provisions in all cases.

If the configuration of a structure is fixed by architectural or other requirements as is normally the case, the structural designer has a restricted choice in the development of the strength and ductility required to ensure adequate seismic behavior. It is often possible to say that some design layouts are better than others for dynamic resistance, although it is fairly clear that different choices of framing can lead to vastly different requirements of strength and ductility. For example, a framed structure is generally less stiff and usually possesses a lower first mode frequency than a shear wall structure with nearly solid walls providing lateral resistance; the design forces may be slightly smaller for the framed structure than for the shear wall structure, and the gross deformation permitted may be larger in the case of the framed structure, although such assessments will depend on the design details.

The methods of analysis employed as part of the design process have implications on the cost and the performance capability of the design. Calculational procedures inconsistent with the assumptions made and parameters used may lead to a false sense of security. If the specifications are unduly conservative the design may be forced into a type that is strong but less ductile, i.e., less able to deform and absorb energy than is desirable. It is difficult to avoid differences in the degree of conservatism among different types of structures, and in some cases it is undesirable to do so. Some materials by their nature, including their variability or lack of adequate control of properties, may require a greater factor of safety against the limit of estimated satisfactory performance than other materials, the properties of which are more accurately determinable and controllable. The margin between incipient failure and complete collapse may differ for different materials and may therefore involve a difference in the factor of safety required in the design. It is desirable, in the development of the basis for a performance criterion, that the designer's approach not be too greatly constrained. For example, it may be unwise to prescribe limits for both strength and ductility in such a way that the balance between the two cannot be adjusted to take account of new material properties or new structural types as they are developed. Ideally some degree of trade-off between ductility and strength should be available in the methods that are permitted, so as to achieve economy without the sacrifice of safety. But whether or not one is interested in achieving strength or ductility, or both, the materials have to be used in an appropriate fashion, and adequate methods of inspection and control of construction are needed to ensure that their use is proper.

The most desirable type of design code or criteria specification is one that puts the least restrictions on the initiative, imagination, and innovation of the designer. Such a code might involve only criteria for: (1) the loading or other seismic effects; and (2) the level of response (for example, stress or deformation) or the performance of the structure under the specified loading or other effects. Such an approach need not, and preferably should not, indicate how the designer is to reach his objective, provided he can demonstrate—through documentation of adequacy or through demonstration of adequacy from past performances—that the building has achieved a structural capability to resist the specified conditions; and, the building official, regulatory reviewer, or even owner within the limits of their powers and knowledge, must be convinced that the design is adequate. This approach, for example, generally is the one used in recent years for the design of nuclear reactor power stations as well as many other types of important facilities. Experience over the past several years in approaching seismic design criteria in this way has indicated some problems, but also appears to have been reasonably successful in avoiding constraints due solely to the specifications themselves, although some constraints normally are imposed that are based on the environmental conditions and the stress and deformation levels permitted. Obviously the final proof of adequacy will occur when the facilities experience an earthquake.

Union Bank Building in Los Angeles. This 42-story, steel-frame structure was the first building to be designed on the basis of dynamic response calculations. The response of the building was calculated on a digital computer for five different accelerogram inputs: ground motion corresponding to a M8 earthquake on the San Andreas fault 35 miles distant, and four other inputs representing smaller earthquakes on nearer, smaller faults.

Earthquake Ground Motions

General Observations

The process of earthquake-resistant design requires selection of the earthquake hazard and associated load and deformation effects, as well as an estimate of the structural resisting strengths, as an integral part of the procedure. Unless these determinations are made in a consistent manner, the final design may be either grossly uneconomical or dangerously unsafe. Both sets of parameters are probabilistic in nature although, for convenience, many aspects of the quantification of structural strength may reasonably be approximated as deterministic. However, the earthquake motions for which the design is to be accomplished, or even the occurrence itself of an earthquake affecting the site, must be considered as probabilistic.

In general, two procedures are available to define the earthquake hazard. In the first, where there is an extensive history of earthquake activity and geologic and tectonic investigations are feasible, estimates often can be made of the possible magnitude and location of future earthquakes affecting a site. In many instances it can be assumed such earthquakes will occur along well defined faults. One can then make estimates of the earthquake motion intensity propagated to the site, taking into account the experimental and observational data available, as well as the great variation in nature of such data. Typical of the many relatively recent studies of this type involving world-wide data sets, and in some cases probabilistic approaches, are those by Schnabel and Seed (Ref. 1), Donovan et al. (Refs. 2, 3), Campbell (Ref. 4), and Idriss (Ref. 5). Recent strong motion acceleration data recorded in the near-field (including, for example, the data from the 1979 Imperial Valley earthquake, the 1979 Coyote Lake earthquake, and others) have provided an augmented data base. As a result of the strong ground motion instrumental data obtained since about 1932 from distances of up to several hundreds of kilometers from the source—over a time period of about 50 years, a short time base for making longterm inferences —numerous studies of the type cited have been undertaken in which the motions have been expressed as a function of the size of the earthquake (normally by some designation of magnitude), type of faulting, distance from the fault or estimated source of energy release, regional and site geology, duration of shaking, and other parameters. Recent seismological studies suggest that some forms of ground motion saturation may be evident in near source regions. In all cases noted above, as well as for the approach that follows, great care and judgment are required in interpreting the results and evaluating the significance of the associated uncertainties.

A second procedure is often used for developing the earthquake hazard in a region where occurrence of earthquakes cannot be clearly identified with specific faulting or when insufficient ground motion data are available, as is the general case in the eastern United States. Under these conditions, relationships have been developed for correlating ground motions, generally maximum velocities or accelerations, to a qualitative measure of the intensity of motion, as for example that of the Modified Mercalli (MM) Intensity. Although these relationships are not as readily subject to mathematical determination as the relationships for earthquake shock (motion) propagation, there are sufficient observations to permit useful probabilistic data to be obtained. Examples of such correlations are summarized by Ambraseys (Ref. 6) and Trifunac and Brady (Ref. 7).

Intensity data exhibit even more scatter than those obtained from accelerations measured at some distance from the focus. They are complicated by the fact that the MM intensity is a subjective measure of many different types of damage in large part, and for higher levels of damage they depend to a great extent on the type and age of the building, properties of building materials, methods of construction, foundation conditions and the like. For these reasons one would expect some changes in damage assessment over scores of years as the quality of construction materials and design practice improved. For the reasons noted, and in light of the reporting techniques employed over the years (e.g., lack of knowledge as to whether or not the MM intensity damage report included ten damaged buildings out of 200 buildings, or out of 200,000 buildings), MM intensity data can be considered as extreme value data in a probabilistic sense and often leads to higher predicted peak accelerations than might reasonably be expected for design purposes.

Several recent statistical studies have been made of vertical and horizontal earthquake motions (Refs. 8, 9 and 10). Although the scatter in results is quite great, it is recommended that the design motions in the vertical direction be taken as 2/3 of the value in the horizontal direction across the entire frequency range. Some limited recent near-field data suggest that in the epicentral region in the high frequency range the vertical excitation may equal the horizontal excitation; often it is noted that the vertical motion occurs quite early in the acceleration time-history record, i.e., not in phase with the strong lateral motion. As more near-field data become available for different geological settings, these phenomena and their importance to structural response need to be examined in more detail.

Site Amplification and Modification by Presence of Structure

The regional motions derived from the methods described must be modified to take account of the geologic and stratigraphic conditions pertaining to the site. Although there has been a great deal of study and research on this topic, it must still be considered a somewhat controversial matter. Nevertheless, it is clear from observations that the type of faulting, the regional geology, the local site conditions and the nature of the structure have a major influence on the motions that are experienced by the structure. Studies of the nature of the motions on sites of different general stiffnesses are summarized in Refs. 11 and 12 in terms of the response spectra associated with the measured records at various sites, and in Ref. 13 for longer period motions. As Nuttli (Ref. 14) has indicated, long period motions are clearly of major significance in the midwestern United States for large earthquakes, and their effects will be most noticeable for structures or systems with periods near those of the ground motion. The variation in intensity of motion with depth beneath the surface is very complex to handle. As of this time there are few data worldwide that directly relate surface motions to motions beneath the surface.

For evaluating the effects of earthquakes on structures, field observations and instrumental measurements suggest that a complex interaction occurs. This important, yet not well understood, subject commonly described by the term "soil-structure interaction" is the topic of a separate monograph in this series. Excellent summaries of the state of the art can be found in Refs. 15 and 16. There are two fundamental aspects to the phenomenon, namely

1. the characteristics of the earthquake ground motions are modified by virtue of the presence of the structure; and

2. the structure interacts with the adjacent and supporting soil materials which in turn affect the building system vibration characteristics.

A comprehensive soil-structure interaction analysis at a site involves (a) evaluation of the soil properties (linear and nonlinear) in the supporting medium, including properties at depths, (b) evaluation of the nature of the wave propagation associated with the ground motions, (c) analysis of the soil-structure system as modelled, and (d) evaluation of possible interaction effects arising from neighboring structures.

Many analysis techniques for estimating soil-structure interaction effects have been developed and can be conveniently classified as falling into two distinct yet broad classes, namely direct methods and multistep or substructure methods. Among the more prominent analysis techniques currently employed are those based on finite element modelling and those centering around the use of impedance functions. Although the problem is very complex and all the computational techniques currently employed have limitations of one type or another, research has led to continuing improvement in the ability to predict interaction effects. However, in view of the fact that the field is a developing one, it is suggested that reliance solely on any single method be avoided and that careful judgment be exercised using theory and observation jointly, to arrive at estimates of interaction effects.

Future theoretical development of soil-structure interaction computational techniques, and verification thereof, will depend in part on obtaining measurements (in arrays) in the free field on the surface and at depth, and on structures, especially at the base level.

Actual Versus Effective Earthquake Motions

It is desirable to begin this discussion with some attention to structural resistance. Well designed and well constructed buildings have survived significant earthquake ground motion, even though most earthquake reconnaissance reports center around descriptions of damaged buildings and equipment. These studies of failures have been quite valuable and have contributed to the upgrading of earthquake engineering practice; however, surprisingly little attention has been devoted to detailed studies of lightly damaged or undamaged buildings and equipment, and the reasons for their survival. As a result, documentation concerning this topic is only fragmentary. Earthquake damage reports suggest that the damage in well engineered buildings often does not correlate well with the observed peak ground acceleration. Typical of citations to this effect are those listed in Refs. 17–24. Our base of knowledge for assessing the adequacy and margins of strength of structures and equipment must be improved in the years ahead through additional research involving studies of structural and mechanical systems that *have not* failed, as well as those that *have* failed.

The instruments normally employed for making free-field ground motion measurements are strong motion accelerographs. Because acceleration is at least in part a measure of the force involved and because the high frequency (nearly zero period) acceleration is commonly used as the anchor point for design response spectra, it is natural that acceleration has been employed so commonly as a descriptor of the ground motion. All three types of ground motion input (acceleration, velocity and displacement) are important and must be considered in an appropriate manner as part of the design process. However, one senses that there has been significant over attention given to acceleration, especially when considered in the context of the manner in which it is employed in the design of structural and mechanical systems. In some respects velocity may be a better descriptor in that it is a measure of the energy involved in the response process, and may well be a better measure for assessing the parameters associated with the response and potential damage of a structure or equipment system. These topics are the subject of considerable research in terms of examining seismic input motion parameters and of developing characteristic motion and response identification parameters and their role in resistance to excitation and structural damage.

Recent research (as reported in Refs. 25 and 26, for example) as well as field observations, suggests that high-frequency spikes of acceleration do not have a significant influence in the response and behavior of mechanical and structural systems and that the repetitive shaking with strong energy content is the characteristic of the time history that leads to structural deformation and damage.

In most cases extremely high accelerations can occur on a localized basis without significant damage to structures or equipment. Many types of structures, as well as equipment, are designed to resist very high acceleration in the range of hundreds of gravities or more, as for example in the case of military structures and equipment. If one strikes a building with a structural wrecking ball, localized damage and high accelerations occur in the region where the ball strikes the building; generally, such localized loading for a well engineered structure does not lead to building collapse or even any type of gross damage.

The concept of effective acceleration, as defined recently by Newmark and Hall in some special design studies, may be stated in the following manner:

> It is that acceleration which is most closely related to structural response and to damage potential of an earthquake. It differs from and is less than the peak free-field ground acceleration. It is a function of the size of the loaded area, the frequency content of the excitation, which in turn depends on the closeness to the source of the earthquake, and to the weight, embedment, damping characteristic, and stiffness of the structure and its foundation.

As employed for design and review analyses of critical facilities, the term "effective acceleration" is associated with the significant part of the ground motion containing repetitive motion portions that possess strong energy content and that produce significant linear and nonlinear deformation; obviously, duration of shaking as well as amplitude and frequency (time) characteristics are among the important parameters to be considered. These portions of the ground motion are of primary importance in evaluating the response and behavior of the structure or equipment elements, and thereby are of importance in design and in assessing damage potential. In this sense, then, in accordance with the definition given above, the effective acceleration normally is not the peak instrumentally recorded high-frequency accelerations commonly found to occur close to the source of seismic energy release, especially for structural foundations of some size or weight. On the other hand, the effective acceleration would be expected to be very close to the peak instrumental acceleration for locations at significant distances from the source, zones where such high frequency acceleration peaks normally are not encountered. Accordingly, for design purposes it is believed that the effective acceleration value should be used in the basic process of arriving at the anchor point for the design response spectrum.

With time it seems certain that velocity and displacement control values will assume increasingly important roles in defining relative spectral shape. Moreover, one can foresee where the concept of so-called "effective motion" will be broadened to include motions throughout the entire range of frequencies. There are statistical and probabilistic aspects involved in effective motion too, in the sense that no two earthquakes exhibit identical acceleration, velocity or displacement time histories or, similarly, identical response spectra, even at locations fairly close to each other; similarly there are uncertainties associated with the magnitude, which is a general measure of this overall energy input. Thus, the concept of effective motion must reflect the fact that for a given site and for possible earthquakes in a nearby region, the various sources of strong motion, the varying attenuation through the ground, the local site characteristics, and the nature and form of the structure all have an influence on the motions that may be transmitted through the foundation.

All of the foregoing factors (and others yet to be identified) must be taken into account in some manner, largely judgmentally at present, in arriving at an assessment of the motion that occurs and its applicability to design in any particular situation. The fact that such assessments must be largely judgmental at present is in part one of the principal problems facing those charged with designing (and gaining regulatory approval of) major facilities today, especially those critical facilities sited in highly seismic and highly populated areas. Systematic approaches with supporting technical documentation are required, and are the subject of ongoing research.

Exxon Hondo Tower. This view shows the Exxon offshore tower prior to being towed to the site. This Hondo tower was installed in 850 ft. of water off the coast of Santa Barbara, California, and at this time was the world's largest offshore tower. The design was based on dynamic analyses using as input the strong ground shaking expected at that location.

Design Response Spectra

Spectrum Concepts

The response spectrum is defined as a graphical relationship of the maximum response of a single-degree-of-freedom elastic system with damping to dynamic motion or forces (Refs. 27-39). The concept of a response spectrum has been recognized for many years in the literature in connection with studies of the response of oscillators to transient disturbances. Increased interest in the subject in the structural engineering field arose during and immediately after World War II in connection with studies of transient response arising from many sources including, among others, blast and shock induced ground motion and earthquake ground motion. The results of these studies can be traced through the literature cited. The studies of transient excitation, as well as harmonic excitation, led to development of the basic principles now employed in constructing response spectra as a function of type of excitation, damping, ductility, and the response quantities under consideration. The many techniques that can be employed for computing the response of single-degree-of-freedom (SDF) systems are presented in other monographs in the series and in the references cited, especially Refs. 27, 28 and 37-39. The purpose of this section is to present an

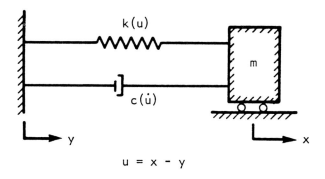

Figure 1. Simple Damped Mass-Spring System.

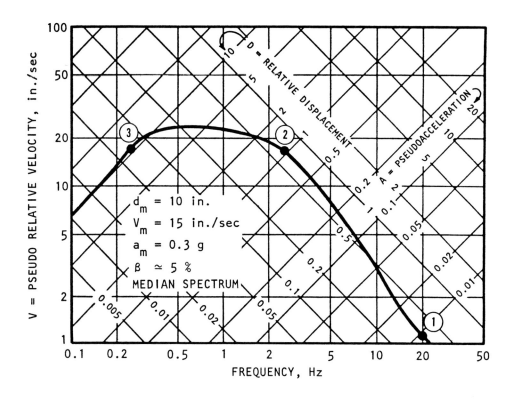

Figure 2. Typical Response Spectrum for Earthquake Motions.

overview of response spectrum concepts for purposes of background, completeness and continuity in the material presented.

In general, it can be shown that the response of a simple damped oscillator (Fig. 1) to a dynamic motion of its base can be represented graphically in a simple fashion by a logarithmic plot as shown in Fig. 2. In this figure, the plot uses four logarithmic scales to show the following three response quantities:

D = maximum relative displacement between the mass of the oscillator and its base
V = maximum pseudo relative velocity = ωD
A = maximum pseudo acceleration of the mass of the oscillator = $\omega^2 D$

In these relations, ω is the circular natural frequency of the oscillator, or $\omega = 2\pi f$ where f is the frequency in cycles per second (cps) or Hertz (Hz).

The effective ground motions for the earthquake disturbance for which the oscillator response of Fig. 2 is drawn in approximate form are maximum ground displacement $d_m = 10$ in., maximum ground velocity $v_m = 15$ in. per sec., and maximum ground acceleration $a_m = 0.3$ g, where g is the acceleration of gravity. The response curve shown is a smooth curve rather than the actual jagged curve that one normally obtains from a precise calculation, and is given for a damping ratio (β) of about 5 percent of critical. The symbols 1, 2 and 3 on the curve denote the responses of different oscillators, Item 1 having a frequency of 20 Hz, Item 2 of 2.5 Hz, and Item 3 of 0.25 Hz. It can be seen that for Item 1 the maximum relative displacement is extremely small, but for Item 3 it is quite large. On the other hand, the pseudo acceleration for Item 3 is relatively small compared with that for Item 2. The pseudo relative velocities for Items 2 and 3 are substantially larger than that for Item 1.

The advantage of using the tripartite logarithmic plot, with frequency also plotted logarithmically, is that one curve can be drawn to represent the three quantities D, V, and A. The pseudo relative velocity is nearly the same as the maximum relative velocity for higher frequencies, essentially equal for intermediate frequencies, but substantially different for very low frequencies. It is, however, a measure of the energy absorbed in the spring. The maximum energy in the spring, neglecting that involved in the damper of the oscillator, is $\frac{1}{2}MV^2$, where M is the mass of the oscillator. There are many other ways of graphically plotting response spectra, so the interested reader should consult the many references cited herein for other possible maximum response representations.

The pseudo acceleration is exactly the same as the maximum acceleration where there is no damping, and for normal levels of damping is practically the same as the maximum acceleration. Thus the force, as determined from the mass times the acceleration, is a good representation of the maximum force in the spring.

There are many strong motion earthquake records available.

One record depicting some of the strongest shaking that has been measured is that for the El Centro Earthquake of May 18, 1940. The response spectra computed for that earthquake for several different amounts of damping are shown in Fig. 3. The oscillatory nature of the response spectra, especially for low amounts of damping, is typical of the nature of response spectra for earthquake motions in general. A replot of Fig. 3 is given in Fig. 4 in a dimensionless form where the scales are given in terms of the maximum ground motion components. In this figure, the ground displacement is denoted by the symbol y, and the subscript m designates a maximum value. Dots over the y indicate differentiation with respect to time.

It can be seen from Fig. 4 that for relatively low frequencies, below something of the order of about 0.05 Hz, the maximum displacement response D is practically equal to the maximum ground displacement. For intermediate frequencies, however, greater than about 0.1 Hz, up to about 0.3 Hz, there is an amplified displacement response, with amplification factors running up to about three or more for low values of the damping factor β.

For high frequencies, over about 20 to 30 Hz or so, the maximum acceleration is practically equal to the maximum ground acceleration. However, for frequencies below about 6 to 8 Hz, ranging down to about 2 Hz, there is nearly a constant amplification of acceleration, with the higher amplification corresponding to the lower values of damping. In the intermediate range between about 0.3 to 2 Hz, there is nearly a constant velocity response, with an amplification over the maximum ground velocity. The amplifications also are greater for the smaller values of the damping factor.

The results shown in Fig. 4 are typical for other inputs, either for other earthquake motions or for simple types of dynamic motion in general.

Although actual response spectra for earthquake motions are quite irregular, as just noted, they have the general shape of a trapezoid or tent when plotted on logarithmic tripartite graph paper. A simplified spectrum is shown in Fig. 5, plotted on a logarithmic tripartite graph, and modified so that the various regions of the spectrum are smoothed to straight line portions.

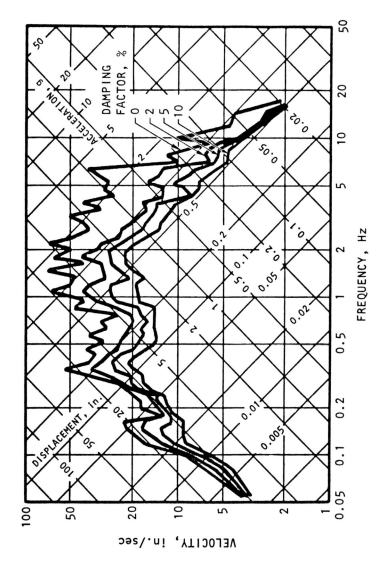

Figure 3. Response Spectra, El Centro Earthquake, May 18, 1940, North-South Direction.

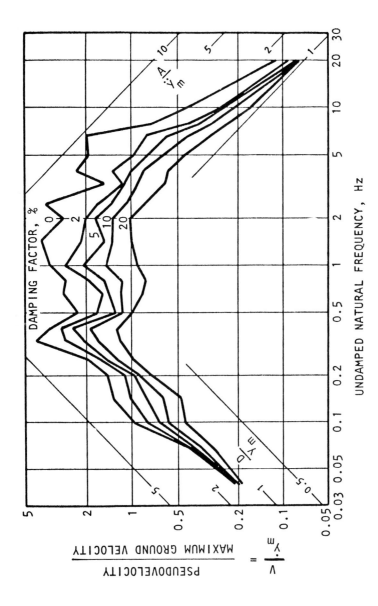

Figure 4. Deformation Spectra for Elastic Systems Subjected to the 1940 El Centro Earthquake.

On the same graph are shown dotted lines that are drawn at the level of the maximum (effective) ground motion values, and the figure therefore indicates the amplifications of maximum ground motions for the various parts of the spectrum. Values of amplification factors for various amounts of damping are shown in Table 1 for two levels of probability considering the variation as lognormal, and in equation form in Table 2; these relationships are based on studies reported in Refs. 8, 9, and 32. Numerical calculation of the response spectrum bounds employing the tabular values cited is given in Fig. 5. Recommended damping values for various materials and structural types are discussed in a later section on damping and ductility.

TABLE 1. SPECTRUM AMPLIFICATION FACTORS FOR HORIZONTAL ELASTIC RESPONSE

Damping, % Critical	One Sigma (84.1%)			Median (50%)		
	A	V	D	A	V	D
0.5	5.10	3.84	3.04	3.68	2.59	2.01
1	4.38	3.38	2.73	3.21	2.31	1.82
2	3.66	2.92	2.42	2.74	2.03	1.63
3	3.24	2.64	2.24	2.46	1.86	1.52
5	2.71	2.30	2.01	2.12	1.65	1.39
7	2.36	2.08	1.85	1.89	1.51	1.29
10	1.99	1.84	1.69	1.64	1.37	1.20
20	1.26	1.37	1.38	1.17	1.08	1.01

TABLE 2. EQUATIONS FOR SPECTRUM AMPLIFICATION FACTORS FOR HORIZONTAL MOTION

Quantity	Cumulative Probability, %	Equation
Acceleration	84.1 (One Sigma)	$4.38 - 1.04 \ln \beta$
Velocity		$3.38 - 0.67 \ln \beta$
Displacement		$2.73 - 0.45 \ln \beta$
Acceleration	50 (Median)	$3.21 - 0.68 \ln \beta$
Velocity		$2.31 - 0.41 \ln \beta$
Displacement		$1.82 - 0.27 \ln \beta$

For the case where limited nonlinear behavior is to be incorporated in the analysis, one possible approach is to employ modified response spectra, i.e., spectra that correspond to specific nonlinear force deformation relationships. Studies suggest that the approach is reasonably accurate for building system analysis when the deformation is limited to ductility values of 5 or 6 or less, as defined next. In order to illustrate the approach, consider the case in which a simple mass-spring oscillator deforms inelastically with a relation of resistance to relative displacement as in Fig. 6. In this figure the elasto-plastic approximation to the actual curvilinear resistance-displacement curve is drawn so that the areas under both curves are the same at the effective elastic displacement u_y and at the selected value of maximum permissible displacement u_m. The ductility factor μ is defined as

$$\mu = u_m / u_y \tag{1}$$

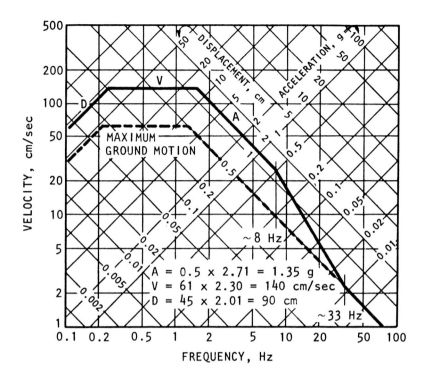

Figure 5. Elastic Design Spectrum, Horizontal Motion (0.5 g Maximum Acceleration, 5 Percent Damping, One Sigma Cumulative Probability).

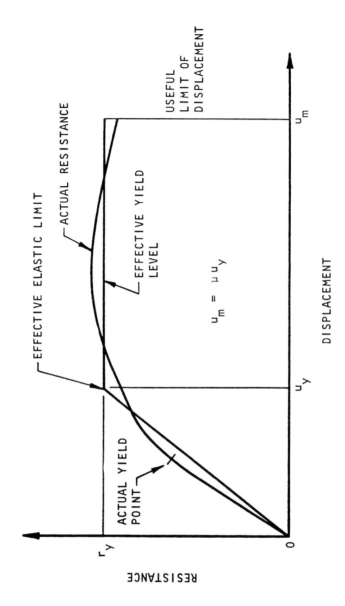

Figure 6. Resistance-Displacement Relationship.

It is convenient to use an elasto-plastic resistance-displacement case because one can draw response spectra for such a case in generally the same way as the spectra were drawn for elastic conditions in Fig. 5. In Fig. 7 there are shown acceleration spectra for elasto-plastic systems having 2 percent of critical damping for the El Centro 1940 earthquake. Here, the symbol D_y represents the elastic component of the response displacement, but is not the total displacement. Hence, the curves give the elastic component of maximum displacement as well as the maximum acceleration A, but they do not give the proper value of maximum pseudo velocity. The figure is drawn for ductility factors ranging from 1 to 10. It is typical of other acceleration spectra for elasto-plastic systems, as indicated by the acceleration spectra shown in Fig. 8 for the step displacement pulse sketched in the figure.

This step displacement pulse corresponds to the two triangular pulses of acceleration shown, where the total length of time required to reach the maximum ground or base displacement is 1 second. The frequency scale shown in Fig. 8 can be changed for any other length of time, t, to reach the maximum displacement by dividing the frequencies f by t. In other words, for a step displacement pulse that takes 0.2 sec, the abscissa for a frequency of 1 Hz would be changed to 5 Hz, and that for 3 Hz in the figure would be changed to 15 Hz, etc. The general nature of the similarity between Figs. 7 and 8 is important and is indicative of the extensive studies of classical pulses which have led to an understanding of the basic response trends, amplification, bounds, etc.

One can also draw a response spectrum for total displacement, as shown in Fig. 9. This figure is drawn for the same conditions as Fig. 7, and is obtained from Fig. 7 by multiplying each curve's ordinates by the value of the ductility factor μ shown on that curve. It can be seen that the maximum total displacement is virtually the same for all ductility factors, actually perhaps decreasing even slightly for the larger ductility factors in the low frequency region, for frequencies below about 2 Hz. Moreover, it appears from Fig. 7 that the maximum acceleration is very nearly the same for frequencies greater than about 20 or 30 Hz for all ductility factors. In between there is a transition. These remarks

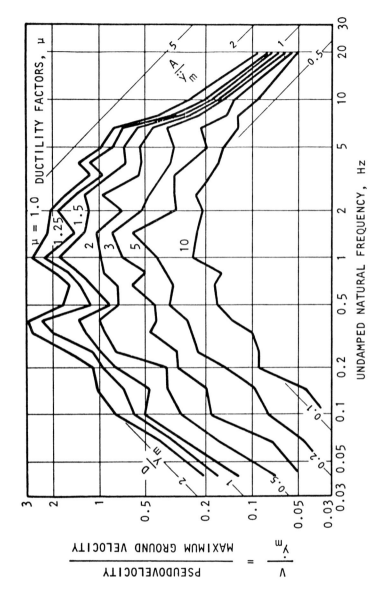

Figure 7. Deformation Spectra, Elastoplastic Systems, 2 Percent Critical Damping, El Centro 1940 Earthquake.

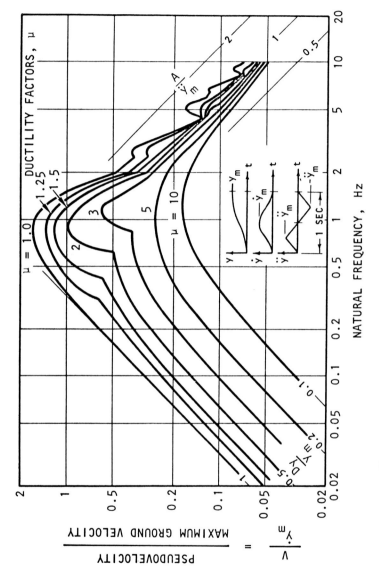

Figure 8. Deformation Spectra for Undamped, Elastoplastic Systems Subjected to a Half-Cycle Velocity Pulse.

are applicable to the spectra for other earthquakes also, and can be generalized in the following way.

For low frequencies, corresponding to about 0.3 Hz as an upper limit, displacements are preserved. As a matter of fact, the inelastic system may have perhaps even a smaller total displacement than elastic systems. For frequencies between about 0.3 to about 2 Hz, the displacements are very nearly the same for all ductility factors. For frequencies between about 2 up to about 8 Hz, the best relationship appears to be to equate the energy in the various curves, or to say that energy is preserved, with a corresponding relationship between deflections and accelerations or forces. There is a transition region between 8 and 33 Hz. Above 33 Hz, the force or acceleration is nearly the same for all ductility ratios. For convenience, one might modify these relationships slightly, as discussed next. The trends just noted are in general agreement with a comprehensive study just completed (Ref. 40) that includes consideration of elasto-plastic, bi-linear and degrading resistance relationships.

For seismic analysis purposes the foregoing relations can be approximated as follows in the general form of response spectra described earlier. Both elastic and inelastic spectra for both acceleration and total displacement are shown in Fig. 10. Here the symbols D, V, A, A_0 refer to the bounds of the elastic response spectrum; the symbols D', V', A', A_o to the bounds of the elasto-plastic spectrum for acceleration or yield displacement; and the symbols D, V, A'', A_o'' to the bounds for the elasto-plastic spectrum for displacement. The symbol A_o refers to the maximum effective ground acceleration.

In general, for small excursions into the inelastic range, when the latter is considered to be approximated by an elasto-plastic resistance curve, the acceleration response spectrum (providing an estimate of the acceleration or yield displacement) is decreased by a factor that is one over the ductility factor, $1/\mu$. Then the reduction for the two left-hand portions (D and V) of the elastic response spectrum shown in Fig. 5 (to the left of a frequency of about 2 Hz) is reduced by the factor $1/\mu$, and by the factor of $1/\sqrt{2\mu-1}$ in the constant acceleration portion (A) to the right, roughly between frequencies of 2 and 8 Hz. There is no reduction beyond about 33 Hz. The response spectrum depicting total

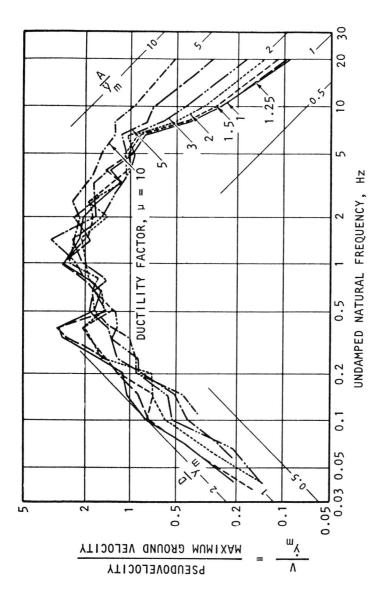

Figure 9. Total Displacement Response Spectra, Elastoplastic Systems, 2 Percent Critical Damping, 1940 El Centro Earthquake.

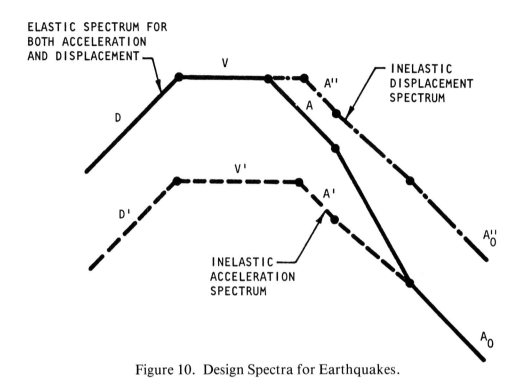

Figure 10. Design Spectra for Earthquakes.

displacement is obtained by multiplying the latter curve by μ over the given frequency range. With this concept, one can arrive at design spectra that take account of inelastic action even in the small ranges of inelastic behavior.

In the light of the preceding discussion, we can now develop a basis for design of structures, elements, or components, where these are subjected directly to the ground or base motion for which we have estimated effective values of maximum displacement, velocity, and acceleration.

In determining the ground motions, it is recommended that, lacking other information, for competent soil conditions a v/a ratio of 48 in./sec/g be used and for rock a v/a ratio of about 36 in./sec/g be used. Also to ensure that the spectrum represents an adequate band (frequency) width to accommodate a possible range of earthquakes it is recommended that ad/v^2 be taken equal to about 6.0. In the above a, v and d are the maximum effective values of horizontal ground motion (acceleration (g), velocity (in./sec), and displacement (in.), respectively).

For the concepts described above and for the selected values of damping and ductility, the design spectrum for earthquake motions can be drawn as shown in Fig. 10 generally. The response spectrum indicated by the line $DVAA_0$ in Fig. 10 is the elastic response spectrum, using the probability levels, damping values and amplification factors appropriate to the particular excitation and structural component. By use of the ductility reductions described, one obtains the design spectrum for acceleration (or force or yield displacement) by the curve $D'V'A'A_0$, total displacement by the curve $DVA''A_0''$, for the elasto-plastic resistance curve. A specific example of the design spectrum that includes all of the quantities described for a peak ground acceleration of about 0.16 g, critical damping of 5 percent and a ductility factor of 3 is presented in Fig. 11.

Modification of Spectra for Large Periods or Low Frequencies

The response spectra of Figs. 5 and 11 have a constant velocity response, V, in the range of frequencies below about 2 Hz, with a cutoff to a constant displacement line at about 0.2 Hz. For structures with long periods, greater than about 1 sec, the spec-

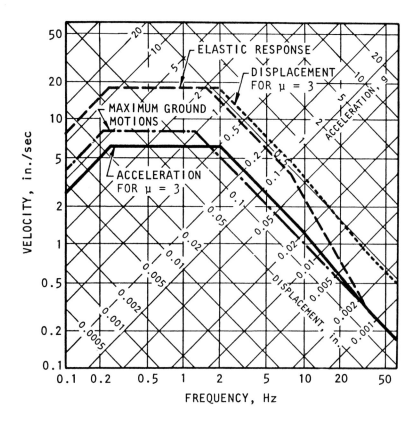

Figure 11. Method of Constructing Elastic and Inelastic Design Spectra.

tral values are not conservative enough for the lower modes of vibration in the modal analysis procedure and definitely not conservative enough for the equivalent lateral force procedure. In order to arrive at a more conservative spectrum for the design, taking account of various uncertainties involved in the combination of modal responses, and possibly other factors such as motions arising from distant earthquakes, the following procedure is suggested. In the range of frequencies below about 1 Hz (or beginning at the knee of the response spectrum), the velocity spectral response value should be taken to vary as the reciprocal of the frequency to the 1/3 power, or as the period to the 1/3 power, instead of being constant as shown in the figures. This corresponds to spectral acceleration values in this range varying as the reciprocal of the period to the 2/3 power, or directly as the frequency to the 2/3 power, instead of varying as the reciprocal of the period or directly as the frequency, as in the figures. In either case, it is suggested that the spectrum be considered to correspond to a constant displacement equal to the amplified ground displacement for periods longer than about 6 sec, or frequencies lower than about 1/6 Hz.

It may be wise to note at this point that as additional knowledge is gained, and especially for the design of special structures or facilities, it may be desirable or necessary to develop a set of response spectra representative of a near-field source, as well as moderate and far distant sources.

Pacoima Dam. This concrete arch dam was located over the center of seismic energy release during the 9 February 1971 San Fernando, California earthquake, M6.4, but sustained no damage. The rock abutment showed some distress but the concrete structure survived intact. The dam was being designed at the time of the 1925 Santa Barbara earthquake and special attention was given to the seismic design. The photograph was taken after the 1971 earthquake at a time of low reservoir level; the reservoir was not full at the time of the earthquake.

Dynamic Structural Analysis Procedures

Introduction

The procedures described herein are restricted essentially to buildings although in some cases they can be applied to other types of structures. In that which follows, some of the factors involved in rigorous procedures of analysis are discussed, and are followed by introductory comments pertaining to approximate procedures of the kind that are normally used in design practice. Many different approaches to analysis are employed in design ranging from purely linear elastic analysis through techniques that account for at least some degree of nonlinear behavior. Most analysis techniques commonly employed in the design of buildings today, including most procedures incorporated in building codes, either implicitly or explicitly include provisions for nonlinear behavior. This section provides a brief treatment of the modal analysis procedure and the equivalent lateral force procedure, and is presented in notation that is employed in the Applied Technology Council provisions (Ref. 41). Special design provisions, i.e., topics that must receive consideration as a part of the design process, are included in the following section. The monograph concludes with a brief overview of the Applied Technology Council (ATC) provisions as recently prepared and issued.

It may be well to note at this point that because the structure is supported on a foundation the interaction of the structure and its foundation is an important factor in determining the structural response. This subject, which is quite complex, is to be treated in a separate monograph devoted to that topic. Some relatively simple and approximate provisions that are applicable in certain instances are presented in the ATC provisions (Ref. 41).

From the most general point of view, analysis of structural response to earthquake motion should consider as many components of ground motion as the number of degrees of freedom of the base: six for a single rigid base, many more for flexible foundations (for example, spread footings without foundation girders), dams, bridges, etc. However, standard strong motion instruments do not record rotational components of ground motion and in the past usually have not been installed in a grid that is dense enough to provide information on variation of ground motion over distances of the order of base dimensions of structures. In order to study their possible effects on structural response, ground rotations and spatial variations in translational motions have been estimated through idealized assumptions about the ground motion. In most analyses, however, no more than the three translational components (two horizontal and one vertical) of ground motion are considered in the design process except in special situations.

A complete three-dimensional analysis of buildings is a formidable task. Most existing computer programs for inelastic, three-dimensional, dynamic analysis of structures do not incorporate an extensive collection of structural elements appropriate for buildings and are therefore of limited application in building analysis. Usually separate planar models in the orthogonal horizontal directions are analyzed for one component of ground motion; numerous computer programs have been developed to implement such analyses. However, planar analyses may not always be reasonable even for a building with nearly coincident centers of mass and resistance because strong coupling between lateral motions in two orthogonal directions and torsional motions can arise if the lower natural frequencies are nearly coincident or if there are asymmetric inelastic effects.

Obviously the results of nonlinear response history analysis of mathematical models of complex buildings would be most reliable if the model is representative of the building vibrating at large amplitudes of motion, that is, large enough to cause significant yielding. Extensive experimental research on force-deformation behavior of structural components at large deformations and shaking table studies on models of complete structures have improved our understanding of inelastic properties of structures. However, it is still difficult to construct mathematical models that lead to satisfactory results and that are not complicated to the point of becoming impractical for analysis of complex structures. Moreover, our understanding of how to provide for distributed or balanced inelastic behavior in a complicated structure, rather than yielding only in the columns, for example, is lacking to some degree, as is our understanding of how to limit nonlinear behavior under repetitive loading. Because dispersion in nonlinear responses to several design ground motions is rather large (Refs. 27 and 28), reliable results can be achieved only by calculating response to several respresentative ground motions—recorded accelerograms or simulated motions—and examining the statistics of response. For the reasons noted, rigorous inelastic response history analyses, especially three-dimensional analyses, would therefore be an impractical requirement in the design of most buildings.

The most commonly used of the simpler methods of analysis are based on the approximation that effects of yielding can be accounted for by linear analysis of the building using the design spectrum for inelastic systems, determined from the elastic design spectrum and allowable ductility factor. (See, for example, Refs. 27-34) Forces and displacements associated with each horizontal component of ground motion are separately determined by analysis of an idealization of the building with one lateral degree of freedom per floor in the direction of the ground motion being considered. Such analysis may be carried out by either the *modal analysis procedure* or a simpler method that will be referred to as the *equivalent lateral force procedure.* Both procedures lead directly to lateral forces in the direction of the ground motion component being considered. The main difference between the two procedures lies in the magnitude and distribution of the lateral forces over the height of the building. In the modal analysis procedure the lateral forces are based on properties of the natural vibration modes of the building, which are determined from the distribution of mass and stiffness over height. In the equivalent lateral force procedure the magnitude of forces is based on an estimate of the fundamental period and on the distribution of forces as given by simple formulas appropriate for regular buildings. Otherwise the two procedures have similar capabilities and are subject to the same limitations.

The direct results from either procedure permit one to ascertain the effects of lateral forces in the direction under consideration: story shears, floor deflections, and story drifts. The story moments also are obtained directly in both procedures, but normally a correction factor is introduced in the equivalent lateral force procedure. Also normally receiving consideration in the two procedures are the following items: effects of the horizontal component of ground motion perpendicular to the direction under consideration in the analysis; torsional motions of the structure; vertical motions of the structure including those arising from horizontal ground motions; and, gravity load effects or the so-called P-Δ effect. Many of these topics are discussed later herein.

A preliminary design of the building must be available before the modal analysis procedure or any of the more rigorous procedures can be implemented, because these procedures require the mass and stiffness properties of the building. The equivalent lateral force procedure is ideally suited for preliminary design, and normally is needed even if more refined analysis procedures are employed in the final design process.

Building Properties and Allowable Ductility Factor for Analysis

Mass and stiffness. The masses to be used in the analysis of a building should be based on the dead loads, plus probable values of the variable or movable live loads. In general it need not be assumed that the design live load for which the floor is designed, for example, should be considered as applied everywhere on each floor at the same time in determining the masses at the various levels in a building subjected to earthquake motions. It is appropriate to use the same proportion of live load that is used as a cumulative factor in the design of the columns in the building. Snow loads can be considered as live loads in the same way, with some average snow load considered as a mass that must be taken into account in the design of the building frame.

In the computation of stiffness of steel members and framing, the behavior of the joints and connections must be considered. This is best taken into account by using center-to-center distances of members between joints, with consideration of the increased stiffness properties near the joints of welded construction, or possibly a decrease for riveted or bolted construction in which the bolts are not high tensile bolts. In reinforced concrete sections, the members are not likely to be completely uncracked, nor is it reasonable to assume that all sections have the reduced stiffness that a cracked section possesses. For this reason, it is appropriate to use an average of the moments of inertia for stiffnesses between cracked and completely uncracked sections or between net and gross sections for reinforced concrete members, unless they clearly are stressed at such low levels that cracking is not likely. Interaction of floor systems with other transverse members such as beams and girders should be considered where the floor system acts as a stiffening element in flexure. In some cases similar considerations may be applicable to walls and floors as well.

The contribution of nonstructural elements, both to mass and to stiffness, should be considered in the design even though these may not be used in developing the required strength parameters of the structure. The added stiffness and mass may contribute to greater moments and shears that must be resisted by the load bearing elements of the structure, but in some cases may provide a degree of redundancy to the structure as well.

Damping and ductility. Damping levels are, as approximated by Coulomb damping, dependent on the level of deformation or strain in a structure. This is reflected in the recommended damping values given in Table 3, where values are suggested in which the percentages of critical damping are given for working stress levels or stress levels no more than one-half the yield point, and for levels of deformation corresponding to stresses at or just below yield levels. In the table the lower values are to be used for structures in which considerable conservatism in the design is desirable, and the upper levels for ordinary structures in general (Ref. 32).

Ductility levels for structures are used in a way that involves a general reduction in the design spectrum. Hence, some reasonable assessment of the allowable ductility factor is required. For this purpose one must be aware of the differences between the various kinds of ductilities involved in the response of structures to dynamic loading. In this respect one must make a distinction between the ductility factor of a member, such as the rotational hinge capacity at a joint in a flexural member, the ductility factor of a floor or story in a building, and the overall ductility

TABLE 3. RECOMMENDED DAMPING VALUES

Stress Level	Type and Condition of Structure	Percentage Critical Damping
Working stress, no more than about ½ yield point	• Vital piping	1 to 2
	• Welded steel, prestressed concrete, well reinforced concrete (only slight cracking)	2 to 3
	• Reinforced concrete with considerable cracking	3 to 5
	• Bolted and/or riveted steel, wood structures with nailed or bolted joints	5 to 7
At or just below yield point	• Vital piping	2 to 3
	• Welded steel, prestressed concrete (without complete loss in prestress)	5 to 7
	• Prestressed concrete with no prestress left	7 to 10
	• Reinforced concrete	7 to 10
	• Bolted and/or riveted steel, wood structures, with bolted joints	10 to 15
	• Wood structures with nailed joints	15 to 20

factor of the building for use in the computation of base shear from the response spectral values.

The ductility factor of a member, or of a floor level or story, and the overall ductility factor of a building are all governed by the development of a resistance-displacement relation, with the displacement being the longitudinal deformation in a tensile or compressive member, the rotation at a joint or connection in a flexural member, or the total shearing deformation in a shear wall. The story ductility factor is essentially defined by use of a relationship in which the displacement is the relative story deflection between the floor above and the floor beneath. The overall system ductility factor is some weighted average in general of the story ductility factors, and is defined best by considering a particular pattern of displacement corresponding to the preferred or executed mode of deformation of the structure.

It can be seen from the discussion that the member ductility factor may be considerably higher than the story ductility factor, which in turn may be somewhat higher than the overall ductility factor. In order to develop an overall ductility factor of 3 to 5 in a structure, the story ductility factors may have to vary between 3 to 8 or 10, and the individual member ductility factors probably lie in the range of 5 to 15 or even more. In this regard it must be remembered that the ductility factor as defined herein is given by the ratio of the maximum permissible deformation to the deformation at the effective yield displacement, rather than the ratio of the maximum deflection to the elastic limit deflection or displacement, as shown in Fig. 6.

Ductility capacities for steel members are generally higher than for reinforced concrete, and for steel structures ductility capacities are higher for tension than for bending, and higher for bending than for compression. Ductility capacities in shear are intermediate between the values in bending and compression, generally. However, the development of high ductilities in flexure or in compression requires that the thickness of outstanding unsupported flanges of members be limited in general to a value of the order of a thickness greater than one-sixth the width of the outstanding leg to develop a ductility factor of the order of 6 or so in compression or on the compression side of flexural members. For special frame details such as eccentric braced forms, great care must be expressed to ensure that the section subjected to heavy shear will function properly; in this case the web, flanges and connecting details must function properly under reversal of loading. The overall system ductility for a steel frame structure may range from 4 to 8.

For reinforced concrete, the ductility capacity is a function of the state of stress and the arrangement of reinforcement. Ductility factors of the order of 10 or more are not difficult to attain in the flexural mode of reinforced concrete beams with equal amounts of compressive and tensile reinforcement. However, without the compressive reinforcement, the ductility factor is less for higher percentages of steel and is inversely proportional to the amount of steel, with a value of the order of 10 being the maximum for 1 percent of tensile steel reinforcement. For members subjected to shear forces, the ductility capacity is also a function of the arrangement and placement of steel, and in general does not exceed values of the order of about 3, and most probably even less for members supporting a high amount of compression such as reinforced concrete columns. However, higher ductility factors can be reached if the concrete subjected to high compression is confined in some manner, such as by spiral reinforcement, in which case ductility factors of the order of 4 to 6 may be reached. For shear walls with diagonal as well as horizontal and vertical reinforcement, and with careful attention to edge reinforcement, ductility factors of 4 to 6 are possible, although lower values of the order of 1.3 to 2 may be applicable for massive reinforced concrete members. In timber, ductility factors of the order of 2 to 4 are possible, and in masonry lower ductility factors, of the order of 1 to 3, may be reached depending upon whether or not the masonry is reinforced. A maximum of about 1.3 is the upper limit for unreinforced masonry.

In a building, the story ductility factor should not vary rapidly from story to story nor should there be a major change in the rate of increase or decrease in story ductility factor with height; by virtue of design for gravity the story design capacity should be based on the relationship of the shear strength provided in that story, relative to the computed story shear. In general, for use of a ductility factor in the range of 5 or 6, it is recommended the smallest ratio of story shear capacity to computed story shear should not be less than 80 percent of the average of these ratios for all stories. For a story ductility of 4, the corresponding ratio need be only about 67 percent.

At this point it is important to distinguish carefully between ductility demand as expressed by the modified response spectrum and the ductility capacity as reflected by the structural resistance relationship. The modified (or reduced) response spectrum, which provides estimates of the maximum acceleration (or force) level or yield deformation as a function of frequency, explicitly assumes that the ductility and damping employed in arriving at the response will be applicable. In other words, if the structural resistance ductility capacity cannot meet the ductility demand assumed for the response spectrum, then the loading estimate from the spectrum is in error. For example, it is possible that the structure may experience larger loadings or unacceptable deformation before reaching the ductility assumed in developing the spectrum response. All of this is intended to emphasize that the designer must have a knowledge of the resistance relationships appropriate for the structure being designed, including nonlinear behavior, as well as the modified loading expressed by the response spectrum. Such considerations are implied in all analysis procedures that employ reduced response coefficients, including those in many building codes. Moreover, it should be noted here that the use of modified response spectra as a basis for design at present is believed to be reasonably accurate for ductilities up to values of 4 to 6, but not for significantly larger values; further research and earthquake field observations and studies may serve to change these suggested limits in the future.

Modal Analysis Procedure

The modal, or mode superposition, method is generally applicable to analysis of dynamic response of complex structures in their linear range of behavior, in particular to analysis of forces and deformations in multistory buildings due to medium intensity ground shaking causing moderately large but essentially linear response of the structure (Refs. 27 and 28). The method is based on the fact that for certain forms of damping—which are reasonable models for many buildings—the response in each natural mode of vibration can be computed independently of the others, and the modal responses can be combined to determine the total response. Each mode responds with its own particular pattern of deformation (the mode shape), with its own frequency (the modal frequency), and with its own modal damping; and the modal response can be computed by analysis of a single-degree-of-freedom (SDF) oscillator with properties (damping and ductility) chosen to be representative of the particular mode and the degree to which it is excited by the earthquake motion. Independent SDF analysis of the response in each natural vibration mode is, of course, a very attractive feature of modal analysis. Even more significant is the fact that, in general, the response need be determined only in the first few modes because response to earthquakes is primarily due to the lower modes of vibration. For buildings, numerous full-scale tests and analyses of recorded motion during earthquakes have shown that the use of modal analysis with SDF viscously damped oscillators describing the response of vibration modes provides a fairly accurate approximation for analysis of linear response.

Strictly speaking, the modal method, which is applicable only for the analysis of linear responding systems, leads to only an approximate estimate of the design forces for buildings because they are usually designed to deform significantly beyond the yield limit during moderate to very intense ground shaking. However, it is believed that for many buildings, satisfactory approximations to the design forces and deformations can be obtained from the modal method by using the modified design response spectrum for inelastic systems (as described earlier herein) instead of the elastic response spectrum. In what follows, the modal method is presented first for elastic systems and later for inelastic systems; the presentation is based on recent ATC design recommendations (Ref. 41).

A complete modal analysis provides the history of response—forces, displacements and deformations—of a structure to a specified ground acceleration history. However, the complete response history is rarely needed for design; the maximum values of response over the duration of the earthquake usually suffice. Because the response in each vibration mode can be modeled by the response of a SDF oscillator, the maximum response in the mode can be directly computed from the earthquake response spectrum, and procedures for combining the modal maxima to obtain estimates (but not the exact value) of the maximum of total response are available and are described later herein.

In its most general form, the modal method for linear response analysis is applicable to arbitrary three-dimensional structural systems. However, for the purpose of designing buildings it often can be simplified from the general case by restricting consideration to lateral motion in a plane. In many cases planar models appropriate for each of two orthogonal lateral directions can be analyzed separately and results of the two analyses combined.

Structural idealization. The mass of the structure normally is lumped at the floor levels; only one degree of freedom—the lateral displacement in the direction for which the structure is being analyzed—per floor is required, resulting in as many degrees of freedom as the number of floors.

Modal periods and shapes. The design method requires knowledge about periods and shapes of vibration for each of those natural modes of vibration that may contribute significantly to the total design quantities. These should be associated with a moderately large but essentially linear response of the structure, and the calculations should include only those building components that are effective at these amplitudes. Such periods may be longer than those obtained from a small-amplitude test of the building, or the response to small earthquake motions, because of the stiffening effects of the nonstructural and architectural components at small amplitudes. During response to strong ground shaking, the measured responses of buildings have shown that the periods lengthen, indicating in part a loss of the stiffness once contributed by these components.

Several methods for calculating natural periods and associated mode shapes of vibration of a structure are available (Refs. 27 and 28). The calculations can be carried out readily by standard computer programs that are widely available.

As mentioned earlier, responses of buildings to earthquake motion can be estimated with sufficient accuracy usually through use of only the first few modes of vibration. For determining design values of forces and deformations, three modes of vibration in each lateral direction are nearly always sufficient for low- and medium-rise buildings, but more modes may be necessary in the case of high-rise buildings; six modes in each direction generally would be sufficient in the latter case.

Modal responses. Maximum responses in each natural mode of vibration can be expressed in terms of the modal properties and the earthquake response spectrum. For the nth mode, the base shear force component α_{on} is:

$$\alpha_{on} = \frac{A_n}{g} \beta_n \qquad (2)$$

where A_n is the ordinate, corresponding to the nth mode of vibration having the natural period T_n and damping ratio ξ_n, of the pseudo acceleration response spectrum; g is the acceleration of gravity; and β_n the effective weight, or portion of weight, of the building that participates in the nth mode. The notation employed in the following sections was selected to facilitate comparison with the Applied Technology Council provisions (Ref. 41) and the presentation in the companion EERI monograph by A. K. Chopra (Ref. 42). (See also Ref. 43)

$$\beta_n = \frac{\left[\sum_{i=1}^{N} w_i \phi_{i\,n}\right]^2}{\sum_{i=1}^{N} w_i \phi_{i\,n}^2} \qquad (3)$$

where w_i is the weight lumped at the ith floor level ($w_i = m_i g$), ϕ_{in} is the modal displacement of the ith floor, and N is the total number of floor levels. Equation 3 gives values of β_n that are independent of the method of mode normalization. The lateral force at the ith floor level in the nth mode of vibration is

$$q_{in} = \alpha_{on} \frac{w_i \phi_{in}}{\sum_{j=1}^{N} w_j \phi_{jn}} \qquad (4)$$

In applying the forces at the various floor levels to the building, their direction is controlled by the algebraic sign of ϕ_{in}. Hence, the modal forces for the fundamental mode will act in the same direction; for the second or higher modes they will change direction as one moves up the structure. The form of Eqs. 2 and 4 is different from that usually employed (see Refs. 27, 28 and 29), and is chosen here to highlight the relationship between the modal analysis procedure and the equivalent lateral force procedure.

The subsequent calculation of internal forces—story shears and story moments—and deflections associated with the lateral forces for each mode does not involve any dynamic analysis. The lateral loads are applied at each floor level, statics is used to calculate the story shears and story moments, and a static deflection analysis can be employed to determine the floor deflections. The latter is not necessary, however, because floor deflections v_{in} arising from the lateral forces associated with a particular mode, Eq. 4, are proportional to the mode shape, and the two are related rather simply and obviously by

$$v_{in} = \frac{1}{\omega_n^2} \frac{g}{w_i} q_{in} \qquad (5)$$

where $\omega_n = 2\pi/T_n$ is the frequency of the nth natural mode of vibration.

The story shears and moments in individual modes are combined to determine their total values. These total values are distributed to the various frames and walls that make up the lateral force resisting system. Whereas it is convenient in all cases and satisfactory for many buildings to defer such distribution until after the modal values of story forces have been combined, the results may be in error for walls and braced frames. For better evaluation of shears and moments at various levels in walls and braced frames, individual modal values for these quantities should be determined by appropriate distribution of the modal values of story shears and moments and combined directly by techniques described later herein.

Deformation quantities similarly should not be determined from the total floor displacements (after combining the modal values), but individual modal values should be determined and combined. For example, the drift Δ_i in story i should be determined by combining the modal values:

$$\Delta_{i\,n} = v_{i\,n} - v_{i-1,n} \tag{6}$$

Total responses. Total responses of an elastic structure are obtained through the superposition of responses in the natural modes of vibration of the structure, and the maximum responses in individual modes of vibration can be determined from the earthquake response spectrum. Because in general the modal maxima r_n do not occur simultaneously during the ground shaking, the direct superposition of modal maxima provides an upper bound to the maximum of total response r, namely

$$r \leq \sum_{n=1}^{N} |r_n| \tag{7}$$

This estimate is often too conservative and is therefore not recommended. A satisfactory estimate of the total response usually can be obtained from the root-sum-square:

$$r \cong \sqrt{\sum r_n^2} \qquad (8)$$

in which, as discussed earlier, only the lower few modes need to be included in the summation.

The maximum value of any response—story shear, story moment, shear and moment at various levels in braced frames and walls, floor displacement, story drift, etc.—can be estimated by combining the modal values for that response in accordance with Eq. 8. The quality of this estimate is generally good for systems with well separated frequencies, a property typically valid for the building idealization adopted here, wherein only the lateral motion in a plane is considered. Improved combination formulas are available for systems lacking this property (Ref. 27).

Application to inelastic systems. The modal analysis procedure described above is strictly valid only for systems in their linearly elastic range of behavior. With the following modifications, it may, however, be employed as an approximate procedure for analysis of nonlinear responses especially if the ductility ratios employed are reasonably low, as discussed earlier. In Eq. 2, A_n, the ordinate of the pseudo acceleration response spectrum for a linearly elastic system, with vibration period T_n and damping ratio ξ_n, should be replaced by the corresponding value for a nonlinear system, with the same period of small amplitude vibration and damping ratio, which is determined from the design response spectrum for the allowable ductility factor. Accordingly,

$$\alpha_{o\,n} = \frac{A_n'}{g} \beta_n \qquad (9)$$

and this value may be used in Eq. 4 to calculate the inertial forces at each floor level. Thereafter, the displacements are calculated from Eq. 5 by μ, the allowable ductility factor, to obtain the total deflection in the nth mode:

$$v_{i\,n} = \mu \frac{1}{\omega_n^2} \frac{g}{W_i} q_{i\,n} \qquad (10)$$

Combined earthquake design responses. Two independent analyses by the modal analysis procedure described above lead to the effects of lateral forces associated with ground motion in two orthogonal directions. The design forces and deformations due to earthquake effects normally are determined by combining the results of these independent analyses, including the effects of torsional motions of the structure, of vertical motions of the structure due to horizontal ground motion, of the vertical component of ground motion, and of the P-Δ effects arising from gravity loads. All of these topics receive additional attention in later sections herein.

Equivalent Lateral Force Procedure

Although the building is idealized in the same manner as in the modal analysis procedure, the equivalent lateral force procedure requires less effort because, except for the fundamental period, the periods and shapes of natural modes of vibration are not needed. The magnitude of lateral forces is based on an estimate of the fundamental period of vibration and the associated force distribution through simple formulas appropriate for buildings with regular distribution of mass and stiffness over height. Situations where results from the equivalent lateral force procedure may not be satisfactory are discussed later. A number of versions of this procedure, differing in detail but based on the same underlying concepts, can be found in various building codes. However, it is important in interpreting the results of each calculation that the analyst carefully review the code provisions to ascertain the force levels employed in the specific code or regulation. The design force levels vary significantly and one must be careful to ensure that the structural resistance and ductility capacity assumed in the earthquake analysis are consistent with the actual structural resistance and ductility capacity as discussed earlier herein. In this connection great care must be exercised when employing working stress approaches and/or load factor expressions and combinations. As will be noted later, one of the advances recommended in the ATC provisions (Ref. 41) is that of employing strength (limit stress) design concepts for all building materials; the goal is to encourage at least some level of consistency in that aspect of the design process.

Planar models appropriate for each of two orthogonal lateral directions are analyzed separately as described next; the results of the two analyses are combined as discussed later.

Fundamental period of vibration. Methods of mechanics cannot be employed to calculate the vibration period of the building before at least a preliminary design is available. Simple formulas that involve only a general description of the building type—e.g., steel moment frame, concrete moment frame, shear wall system, braced frame, etc.—and overall dimensions such as height and plan size are therefore necessary to estimate the vibration period so that the base shear can be computed to aid in arriving at the initial design. Because pseudo acceleration values in design spectra for inelastic systems with moderate to large values of allowable ductility factor generally decrease with increasing values of vibration period, it is desirable to underestimate the fundamental period so that the computed base shear is conservative.

A formula for estimating the fundamental period for moment-resisting frame buildings is recommended in the ATC Provisions, namely

$$T = C_T H^{3/4} \qquad (11)$$

where $C_T = 0.035$ and 0.025 for steel and concrete frames, respectively, and H is the height in feet of the building. A commonly used formula for reinforced concrete shear-wall buildings and braced steel frames is the following:

$$T = \frac{0.05H}{\sqrt{L}} \qquad (12)$$

where L is the plan dimension in feet in the direction of analysis.

The fundamental vibration period of exceptionally stiff or light buildings may be significantly shorter than the estimate provided by formulas such as Eqs. 11 and 12. Especially for such buildings, and to be used as a general check, the period for an initial design of the building should be computed by established methods of mechanics. An approximate formula, based on Rayleigh's method, is especially convenient, namely

$$T = 2\pi \sqrt{\frac{\sum_{i=1}^{N} w_i v_i^2}{g \sum_{i=1}^{N} q_i v_i}} \qquad (13)$$

in which the v_i (i = 1, 2, . . . N) quantities are the static lateral displacements at the various floor levels, computed on a linear elastic basis, due to a set of lateral forces q_i. Any reasonable distribution for q_i may be selected but it is convenient to use the lateral forces computed from the empirical estimate or the fundamental period as described later herein.

A word of caution at this point may be appropriate. Drift limits as given in codes or specifications should not be used for estimating periods, and normally will lead to periods that are entirely too large. The drift limits are not uniformly achieved by the structural system during excitation, normally include some significant nonlinear component, and are the same generally for all seismic zones. From the latter, one can deduce that if the drifts were used for period computation, the period would be found to be a function of the zone or applicable seismic coefficient, which is nonsensical. The period is a function of the mass and stiffness of the physical structure and its support.

Nonstructural elements participate in the behavior (and reserve strength) of the structure even though the designer may not wish to rely on them for contributing strength or stiffness to the structure under significant excitation. To ignore them would lead to longer periods and usually smaller design forces; hence, they should be considered in calculating the period. However, if subjected to significant deformation, these elements may lose much of their stiffness but conversely may provide additional damping. The role of such elements, as well as normal floors and walls, needs careful consideration as a part of the structural modelling and response evaluation.

Lateral forces. The distribution of lateral forces over the height of a building is generally quite complex because a number of natural modes of vibration contribute significantly to these forces. The contributions of the various vibration modes to the lateral forces and to the base shear depend on a number of factors, including shape of the design response spectrum and natural vibration periods and mode shapes which in turn depend on the mass and stiffness properties of the building. However, these forces are in large part a function of the first (fundamental) mode of vibration. Thus, in the equivalent lateral force procedure, they are determined from formulas similar to those for the first mode, using Eqs. 2, 3, and 4, but modified to account approximately for the effects of the higher modes.

The following formulas for α_o, the base shear, and f_i, the lateral force at each floor i, have been recommended (Ref. 41):

$$\alpha_o = \frac{A'_1}{g} W \qquad (14)$$

in which A'_1 is the pseudo acceleration corresponding to the estimated fundamental period and the appropriate damping, determined from the design response spectrum for the selected damping and ductility factor, and

$$W = \sum_{i=1}^{N} w_i$$ is the total weight of the building; thus,

$$f_i = \alpha_o \frac{w_i h_i^k}{\sum_{j=i}^{N} w_j h_j^k} \qquad (15)$$

in which h_i is the height of the ith floor above the base and k is a coefficient related to the estimated fundamental vibration period as follows:

$$k = \begin{cases} 1 & T \leq 0.5 \text{ sec} \\ (T + 1.5)/2 & 0.5 < T < 2.5 \text{ sec} \\ 2 & T \geq 2.5 \text{ sec} \end{cases} \qquad (16)$$

It is important to understand the basis for these formulas. If β_1, the effective weight for the first mode, is replaced with W, the total weight, Eq. 9 with n = 1 will become identical to Eq. 14 provided the same value of the fundamental period is used to determine A_1 in both cases. The term β_1 will always be smaller than W; typical values for β_1 are between 60 to 80 percent of W, depending on the distribution of weight over the height and the shape of the first mode. Equation 14 therefore would provide a value for the base shear that will be significantly larger than the first mode value; thus it indirectly and approximately accounts for the contributions of the higher modes of vibration.

If α_{01} is replaced with the total base shear α_0, Eq. 4 will become identical with Eq. 15 with k = 1 if the first mode shape is a straight line. For k = 2 the first mode shape is a parabola with its vertex at the base. Equation 15 with k = 1 is appropriate for buildings with a fundamental period of 0.5 sec or less, because the influence of vibration modes higher than the fundamental mode is small in earthquake responses of short-period buildings, and the fundamental vibration mode of regular buildings departs little from a straight line. Although earthquake responses of long-period buildings are primarily due to the fundamental mode of vibration, the influence of higher modes can be significant, and the fundamental mode lies between a straight line and a parabola with its vertex at the base. The force distribution of Eq. 15 with $k = 2$ is therefore appropriate for buildings with a fundamental period of 2.5 sec or longer. Linear variation of k between values of 1 at a period of 0.5 sec, and 2 at 2.5 sec provides a simple transition between the two extreme values.

Story forces. Story shears are related to the lateral forces by equations of statics. The shear in any story is simply the sum of the lateral forces at floor levels above that story. The story moments can be similarly determined from the lateral forces and heights of various stories by methods of statics. However, to obtain design values, the statically computed overturning moments are reduced for the following reasons:

1. The distribution of story shears over height computed from the lateral forces of Eq. 15 is intended to provide an envelope; because of the contributions of several modes, shears in all stories do not attain their maxima simultaneously during an earthquake. Thus the story moments statically consistent with the envelope of story shears will consist of overestimates.

2. It is intended that the design shear envelope based on the simple distribution of lateral forces of Eq. 15 be conservative. If the shear in some story is close to the exact value, the shears in almost all other stories are likely to be overestimated. Hence, the story moments statically consistent with the design shears may well be overestimates.

3. Under the action of overturning moments, one edge of the foundation may lift from the ground, or shift, for short

durations of time. Such behavior tends to lead to a reduction in the seismic forces and consequently the overturning moments.

Up to a 20 percent reduction is, in general, reasonable for the story moments computed statically from the envelope of story shears, based on results of dynamic analysis studies that take into account the first two of the foregoing reasons; but no reduction should be permitted for the upper stories of a building (see Ref. 41). Rotational inertia due to axial deformations of columns and/or base rotation, in turn caused by soil-structure interaction or rotational components of ground motion, increases moments considerably, near the top. In any case, there is hardly any benefit in reducing the story moments near the top of buildings, because design of vertical elements near the top is rarely governed by these moments. Consequently the following values have been recommended for the reduction factor by which the statically computed story moments should be multiplied: 1.0 for the top 10 stories; between 1.0 and 0.8 for the next 10 stories from the top, linearly varying with height; and 0.8 for the remaining stories.

Formerly many building codes and design recommendations, including the 1968 SEAOC recommendations (Ref. 44), allowed large reduction in overturning moments relative to their values statically consistent with the story shears. These reductions appeared to be excessive in light of the damage to buildings during the 1967 Caracas earthquake, where a number of column failures appeared to be due primarily to the effects of overturning moment. In later (1975) versions of the SEAOC recommendations (Ref. 45), no reduction was allowed. However, making no reduction at all appears to be too conservative in light of the reasons mentioned above and the results of dynamic analysis.

Methods for distributing story shears and moments to the various frames and walls that make up the lateral force resisting system are discussed later herein.

Deflections and drifts. A static deflection analysis that assumes linear behavior of the building will lead to a set of floor deflections. In order to account for the inelastic effects, these deflections should be multiplied by the allowable ductility factor that was used in establishing the design spectrum, resulting in the total deflections. The drift in a story is computed as the difference of the deflections of the floors at the top and bottom of the story under consideration.

The deflections and drifts should be estimated as accurately as possible by considering the effects of joint rotation of members, and shear, flexural and axial deformations that occur between floors. The estimate of such deformations can be critical to design. For example, the satisfactory performance of interior partitions, building cladding and other architectural features can be largely dependent on knowledge of the expected deformation pattern; other mechanical properties of the connectors need consideration as well—for example, potential for corrosion, fatigue properties as affected by wind and thermal effects, etc. Likewise, the design of expansion joints, clearances between structures, and interconnecting pipes and conduits will depend in large measure on a knowledge of the relative deformation. Soil-structure interaction also may increase the total drift. In summary, this aspect of the design process requires special attention.

Earthquake design responses. Effects of lateral forces associated with ground motion in two orthogonal directions can be determined from two independent analyses by the equivalent lateral force procedure presented above. The design forces and deformations due to earthquake effects are determined by combining the results of these independent analyses, to include the other effects mentioned, as discussed in the next section.

San Onofre Nuclear Power Plant Unit 2. Massive, reinforced concrete containment structure of the 1100 megawatt unit of the Southern California Edison Company nuclear power plant. This structure and its contents were designed for earthquakes on the basis of a dynamic analysis, taking into account the properties of the foundation material as well as the properties of the structure itself.

(Courtesy of Southern California Edison Company)

Special Design Considerations

Torsion

Torsional responses in structures arise from two sources: (1) eccentricity in the mass and stiffness distributions in the structure, causing a torsional response coupled with translational response; and (2) torsion arising from accidental causes, including uncertainties in the masses and stiffnesses, the differences in coupling of the structural foundation with the supporting earth or rock beneath, and wave propagation effects in the earthquake motions that give a torsional input to the ground as well as torsional motions in the earth itself during the earthquake.

In general, the torsion arising from eccentric distributions of mass and stiffness can be taken into account by determining the distance between the center of mass and the center of shear stiffness, and ascribing an incremental torsional moment in each story corresponding to the shear in that story multiplied by this eccentricity. A precise evaluation of the torsional response is quite complicated because it is necessary to make essentially a three-dimensional response calculation, taking into account the coupled modes of response of the entire structure. However, one can approximate the response by summing from the top story the incremental torsional moments computed as described above, to obtain the total torsional moment in each story.

The static equivalent torsional responses in each story are then determined by computing the twist in each story obtained by dividing the total torsional story moment by the story rotational stiffness. These twists can then be added from the base upward to estimate the total twist at each floor level.

Since these are essentially static responses, they should be amplified for dynamic response using the response spectrum amplification factor. It is probably adequate to use the factor corresponding to the response spectrum amplification factors in Table 1 for the fundamental torsional frequency of the structure. It should be noted that in many design codes no amplification whatsoever is suggested.

Accidental torsion may arise in various ways, for example from unaccounted eccentricities in the structures and nonuniform excitation of the base of the structure. One can take a value of accidental eccentricity of the order of about 5 to 10 percent of the width of the structure in the direction of motion considered to account for the accidental torsional response. Most current building codes use a value of 5 percent. In evaluating the effect of accidental torsion, one should consider the accidental torsion as an increase and also as a decrease in the eccentricity corresponding to the distance between the centers of mass and resistances in the various stories in order to obtain two bounding values. The accidental torsional effects should be computed in the same way as the real torsion (often referred to as calculated torsion) described above. Both the individual effects (accidental and calculated) and the total effect should be considered in the design.

The foregoing discussion pertains primarily to elastic or quasi-elastic effects. For cases of significant inelastic response with accompanying inelastic torsional response and possibly degradation of torsional resistance, the situation can become quite complex. The behavior of nonstructural components also can materially affect the torsional response; initially with small torsional motion the stiffness of these elements may serve to restrict torsional motion, but as they deform or fail, the torsional properties of the structural system will change and significant torsional response may ensue. The foregoing situations are not easily handled analytically and must receive special attention.

Distribution of Shears

The story shears arising from translational and from torsional responses are distributed over the height of the building in proportion to the stiffnesses of the various elements in the building, with the translational shears being affected by the translational stiffnesses of the building and the torsional shears affected by the rotational stiffnesses of the building. The computed stiffness of the structure should take into account the stiffness of the floors and the floor structure acting as a diaphragm or distributing element. In most cases the floor diaphragm can be considered infinitely stiff in its own plane and only the story stiffnesses are of importance. However, if the floor diaphragm is flexible and deforms greatly in its own plane, the distribution of the forces becomes more nearly uniform than that described by the method discussed above. A simplified approach is possible by considering the relative displacements of the building due to translation, and those due to rotation of each story separately, as affected by the diaphragm or floor stiffness, with the stiffnesses being determined by the forces corresponding to a unit displacement in either translation or torsion, respectively. Then one distributes the shears due to translation or rotation in proportion to these stiffnesses.

This basic shear distribution has been assumed and employed in building codes and in actual designs in many cases over the years. For low- and medium-rise buildings with simple framing or structural units, the procedures described can be expected to provide reasonable design bases. Modern computer analyses of building systems wherein flexural, shear, axial and torsional effects can be treated systematically reveal that the so-called shear building concept, as described above, is inappropriate for some classes of buildings, especially many high-rise buildings. The level of appropriate analysis sophistication requires consideration as a part of the design process.

Base or Overturning Moments

The flexural moment about a horizontal axis at the base is of importance in connection with foundation design. The corresponding flexural moments at each floor level (including the lowest floors) are important in connection with the calculation of vertical stresses in the columns and walls of the structure. These moments can be computed from modal analyses or from an equivalent lateral force analysis. Modifications in the base flexural overturning moment can be made by use of the reduction factors given previously herein when the equivalent lateral force method is used. In the case of the ATC procedures the modal analysis method takes account of these effects directly in the structure, although an additional reduction of 10 percent of the moment may be allowed in the modal analysis for the foundation forces only. An additional reduction of approximately 10 percent may be allowed in the case of the equivalent lateral force analysis as discussed in Ref. 41.

Vertical Component of Ground Motion

In highly seismic zones where the vertical acceleration of the ground may be large compared with the acceleration of gravity, the normal procedure of neglecting vertical accelerations in design may not be appropriate. Generally, the bases for neglecting these vertical accelerations are that the building design allows for gravity effects and usually provides for a high factor of safety in the vertical direction, and that the vertical motions may be significantly out of phase with the horizontal motions. The dynamic vertical response can be estimated by use of a vertical response spectrum to obtain the equivalent vertical forces for which the building must be designed. In making this computation, the response spectrum used for horizontal motion may be used as a basis, but with a scale factor of approximately two-thirds, which is generally the maximum effective ratio between vertical and horizontal accelerations in most ground movements due to earthquakes.

Elements particularly vulnerable to vertical components of ground motion are columns and walls in compression, beams or other horizontal elements, and cantilevered elements, where the amplification factors for vertical response may be fairly large or where there is a relatively small factor of safety for reversed or upward accelerations. In these instances, the amplified vertical acceleration can be estimated from the response spectrum and from the period or the frequency of the particular element considered. Where responding items (equipment) are attached or supported by flexible elements, the additional amplification and accompanying force and deformation levels need to be estimated.

Combined Effects of Horizontal and Vertical Motions

Since the building responds in both horizontal directions at the same time and the stresses are caused by both motion inputs, as well as by the simultaneous vertical motion, it is necessary to consider the combined effects of the various directions of input earthquake motion. If these motions are computed individually, one may make the combination in one of several ways. For example, let us define the effect at a particular point in a particular element, such as stress, moment, shear, etc., arising from the horizontal earthquake in one direction as F_1 and from the earthquake in the transverse horizontal direction as F_2. Let us also define the same effect arising from the vertical component of ground motion as F_3, in those instances where this cannot be neglected. In general, the combined specific effect may be computed as the square root of the sum of the squares of the individual effects, so the resultant effect F is given by the relation

$$F = \sqrt{F_1^2 + F_2^2 + F_3^2} \qquad (17)$$

When it is appropriate to neglect the vertical acceleration effect, F_3 may be taken as zero in Eq. 17.

For a general approximation, one may use instead the relationship that the effects of the different directions of motion are given approximately by assuming 1.0 times the effect of input motion in one direction, combined with 1/3 of the effects in the other two directions. In order to use this relationship properly, one must consider the maximum effect as being the one in which the factor of 1.0 is used, or alternatively one should take 1.0 times the inputs for each direction in turn, combined with 1/3 of the inputs for the other two directions. The summation is made for the absolute values of the responses. It can be shown that this relationship gives results that are close to that given by Eq. 17. (See also Ref. 32)

Effects of Gravity Loads

The effect of gravity loads, when a structure is deflected as a result of the transmitted horizontal motions, is to add a secondary moment owing to the eccentricity in the vertical direction of the gravity loads acting through the lateral deflections corresponding to the horizontal responses. As a first approximation, one may compute the horizontal displacements and the effects of the moments produced by the gravity loads directly. In making this calculation, one should use the total horizontal deflection, including the inelastic portion of it, rather than just the elastic component of the horizontal deflection. Consequently, the computed elastic deflection from the design shears and moments computed by the modal analysis or the equivalent lateral force procedure with a reduced design spectrum, must be increased by the ductility factor μ in order to obtain the total deflections and the corresponding total gravity load moments, or P-Δ effects. A quick estimate of these moments can be made to determine whether or not the P-Δ effect is important. In most cases the first iteration normally will be found to be sufficient.

In some cases the P-Δ effect may be judged to be important; in such case the above method underestimates the actual displacements because of the additional displacement caused by the additional moments accompanying the increment in deflection from the first step in the calculation described above. In effect, there is a series of corrections to be added, which requires the calculation of the successive increments in deflection caused by the P-Δ effect and then the additional moments corresponding thereto in successive stages until convergence is reached. One can arrive at a good approximation to this summation by considering a quantity Θ as defining the relative increment in moment, stress or deflection, due to the first step in the P-Δ calculation, and then computing the final drift by use of a factor $1/(1-\Theta)$—to be taken equal to or greater than one—as a multiplier times the effect computed without consideration of the P-Δ effect. (For further details and definition of Θ, see Ref. 41.)

As in previous cases, a word of caution is in order. In some codes or specifications the computed drift corresponds to working stress design conditions or some other stress condition. In such cases the analyst must be careful to estimate the total drift associated with the design earthquake excitation and employ that drift in calculations following along the lines of the noted general procedures.

Limitation and Choice of Lateral Force and Modal Analysis Methods

Among the most important assumptions common to the equivalent lateral force procedure and the modal analysis procedure are the following: (1) forces and deformation can be determined by combining the results of independent analyses of a planar idealization of the building for each horizontal component of ground motion, including torsional moments determined on an indirect, empirical basis; and (2) nonlinear structural response can be determined to an acceptable degree of accuracy by linear analysis of the building by employing a modified design response spectrum for inelastic systems. Both analysis procedures are likely to be inadequate if the dynamic response behavior of the building is quite different from what is implied by these assumptions.

In particular, both methods may be inadequate if the lateral motions in two orthogonal directions and the torsional motions are strongly coupled. Buildings with large eccentricities of the centers of story resistance relative to the centers of floor mass, or buildings with close values of natural frequencies of the lower modes and essentially coincident centers of mass resistance, exhibit coupled lateral-torsional motions. For such buildings, independent analyses for the two lateral directions may not suffice, and at least three degrees of freedom per floor—two translational motions and one torsional—should be included in the idealized model. The modal method, with appropriate generalizations of the concepts involved, can be applied to analysis of the model. Because natural modes of vibration will show a combination of translational and torsional motions, in order to determine the modal maxima it is necessary to account for the facts that a given mode might be excited by both horizontal components of ground motion and that modes that are primarily torsional can be excited by translational components of ground motion. Because natural frequencies of a building with coupled lateral-torsional motions can be rather close to each other, in such cases the modal maxima should not be combined in accordance with the square-root-of-the-sum-of-the-squares formula;

instead special summation rules should be employed (See Refs. 27, 28).

The manner of combining the maximum responses due to the two horizontal components of ground motion depends on the correlation, if any, between these motions. For earthquakes of practical interest, the intensities in all horizontal directions are comparable, and studies suggest that the maxima are only slightly correlated in phasing. Accordingly, the combined response of both components can be estimated as the square root of the sum of squares of the responses to the two individual components.

The equivalent lateral force procedure and both versions of the modal method—the simpler version and the general version with the three degrees of freedom per floor mentioned in the foregoing paragraphs—are all best applicable to analysis of buildings in which ductility demands imposed by earthquakes are expected to be essentially uniformly distributed over the various stories. For such buildings, the maximum ductility allowed for a particular structural system and material may be used in determining the inelastic response spectrum. If ductility demands are expected to be considerably different from one story to the next, a simple approach is to decrease the allowable ductility factors in establishing the inelastic response spectrum, resulting in larger design forces. Whereas this simple approach is a step in the right direction, the foregoing analysis procedures may still err systematically on the unsafe side if the ductility demands are concentrated in a few stories of the building. For such buildings the actual strength properties can be explicitly considered and the distribution of ductility demand determined by nonlinear response analysis. However, nonlinear numerical analyses may not always be practical nor lead to reliable results. However the analyses are made, there needs to be a careful assessment of the remaining margin of safety inherent in the design (or resistance) for the selected hazard (loading), including consideration of the combinations of all applicable loadings, including seismic loadings.

Both procedures are based on the same basic assumptions and are applicable to buildings whose dynamic response behavior is in reasonable conformity with the implications of these assumptions. The main difference between the two procedures lies in the magnitude of the base shear and distribution of the lateral forces. Whereas in the modal method the force calculations are based on computed periods and mode shapes of several modes of vibration, in the equivalent lateral force method they are based on an estimate of the fundamental period and simple formulas for distribution of forces that are appropriate for buildings with regular distribution of mass and stiffness over height. In the following, a criterion to decide whether or not equivalent lateral force procedure will be adequate in a particular situation is presented.

It would, in general, be adequate to use the equivalent lateral force procedure for buildings with the following properties: the seismic force resisting system has the same configuration in all stories and in all floors; floor masses and stiffnesses do not differ by more than about 30 percent in adjacent floors; and cross-sectional areas and moments of inertia of structural members do not differ by more than about 30 percent in adjacent stories. For other buildings, the following sequence of steps may be employed to decide whether or not the modal analysis procedure should be used:

1. Compute lateral forces and story shears using the equivalent lateral force procedure.

2. Dimension structural members approximately.

3. Compute lateral displacements of the structure as designed in Step 2 due to lateral forces from Step 1.

4. Compute new sets of lateral forces and story shears by replacing h^k in Eq. 15 with the displacements computed in Step 3.

5. If at any story the recomputed story shear (Step 4) differs from the corresponding original value (Step 1) by more than 30 percent, the structure should be analyzed by the modal analysis procedure. If the difference is less than this value the modal analysis procedure is unnecessary, and the structure should be designed using the story shears obtained in Step 4; they represent an improvement over the results of Step 1.

This method for determining whether or not modal analysis should be used is efficient as well as relatively effective.

The seismicity of the area and the potential hazard due to failure of the building should also be considered in deciding whether the equivalent lateral force procedure is adequate. For example, even irregular buildings that may require modal analysis according to the criterion described, can be analyzed approximately by the equivalent lateral force procedure if they are not located in highly seismic areas and do not house critical facilities necessary for post-disaster recovery or for protection of people generally. In such cases, great care is required in interpreting the results of the analysis and in designing the structural resisting systems to ensure satisfactory performance during an earthquake.

And finally, as noted earlier, there are clearly many cases where modern advanced computational analysis of the building system is needed. Such a possible requirement needs to be examined as a part of the design process.

Anchorage Airport Control Tower. The six-story, reinforced concrete frame, control tower at the Anchorage airport collapsed during the M8.4 Alaska earthquake of 27 March 1964. It was reported that since the airport was located outside the city limits it was not covered by the building code and, presumably, earthquake forces were not considered in the design.

Applied Technology Council Provisions

Background

To complete this presentation on seismic spectra and design, a brief review of the general provision of the Applied Technology Council Tentative Provisions (Ref. 41) is presented below. These provisions, currently under review and testing, in all likelihood will form the basic background of future building code provisions.

Seismic design codes in the United States were initiated in the late 1920's with some relatively simple equivalent static formulas. The development of earthquake code provisions proceeded somewhat intermittently until the Structural Engineers Association of California (SEAOC) in 1959-60 published its *Recommended Lateral Force Requirements and Commentary*, which was applicable to California seismic conditions. The SEAOC provisions recognized that the seismic forces induced in a structure are related to the structure's mode of deformation and fundamental period. Seismic codes in the United States and in many other countries have since been patterned after the SEAOC provisions. A brief history of seismic codes in the United States is given in Table 4.

TABLE 4. SEISMIC DESIGN CODES IN THE UNITED STATES

1906 San Francisco Rebuilt to 30 PSF Wind

1927 Uniform Building Code ($C = 0.075$ to 0.10)

1933 Los Angeles City Code ($C = 0.08$)

1943 Los Angeles City Code ($C = \dfrac{0.60}{N + 4.5}$, $N \leq 13$ Stories)

1952 ASCE-SEAONC ($C = \dfrac{K_1}{T}$, $K_1 = 0.015$ to 0.025)

1959 SEAOC ($V = KCW$, $C = \dfrac{0.05}{\sqrt[3]{T}}$)

1974 SEAOC
1976 UBC $\Big\} V = ZIKCSW$

1978 ATC-3 Tentative Recommendations

The need for a coordinated effort to review existing requirements and state of knowledge and to develop comprehensive seismic design provisions applicable to all of the country was recognized several years ago. It was realized by design professionals and governmental representatives that the effort would take many years to complete if performed by volunteer committees as codes have been developed in the past. After numerous detailed discussions, the Applied Technology Council started development in November 1974 of comprehensive seismic design provisions for buildings, which can be adopted by jurisdictions throughout the United States.

The ATC-3 project (as it is designated) was conducted under a contract with the National Bureau of Standards (NBS) with funding provided by the National Science Foundation and NBS. The program was a major joint undertaking of the engineering profession and professional community including architects and earthquake engineers, building regulatory officials, code promulgating officials, the research community, and the federal government. The project was part of the Cooperative Federal Program in Building Practices for Disaster Mitigation initiated in the spring of 1972 under the leadership of NBS.

The 85 participants involved in the development of the provisions were organized into five working groups composed of 14 task committees, two advisory groups, one independent review group, and a project executive panel. Two working drafts were submitted to outside review; the February 1976 draft and the January 1977 draft were reviewed by several hundred reviewers. The reviewers represented practice, industry, professional organizations, governmental agencies, code promulgating groups, and universities. The numerous comments received for each draft were reviewed by the various task committees, and changes and clarifications were made as deemed appropriate in the final report. Subsequently the draft tentative provisions have undergone review by the Building Seismic Safety Council.

It is intended that the provisions can be used by model code groups and governmental agencies to develop seismic design regulations for buildings. However, the participants have recommended that the provisions be tested before they are placed in the code promulgating process. In the near term it is planned that a program will be undertaken to test the provisions by having practicing professionals make comparative designs of various types of buildings in different areas of the United States.

Basic Concepts of ATC-3 Provisions

The primary basis for development of the seismic design provisions for buildings was to protect life safety and to ensure continued functioning of essential facilities needed during and after a catastrophe. It was realized that zero risk is not realistically possible or feasible; expenditures to obtain absolute safety (if it were possible) may not be desirable, as the resources to construct buildings are limited and society must decide how it will allocate available resources. In the development of the U.S. seismic codes currently in use, it was recognized that the specified design coefficients and forces are considerably smaller than those that might be encountered in moderate or major earthquakes. The reasons that building specifications employ these lower coefficients are many, and are based in part on the following factors. Normally it is not feasible economically to design all structures to resist in an elastic (linear) manner the high seismic motions that might occur only once in 500 or 5,000 years. Thus, consideration of the low likelihood of experiencing such motions enters into a determination of a deemed acceptable risk and thus the rationale for being satisfied that a lower seismic coefficient is generally acceptable, especially from an economic and risk point of view as noted above. Obviously, personal safety is considered strongly in arriving at the acceptable risk. Past investigations have studied the frequency-dependent responses and the behavior and damage mechanisms of many types of structures, and as a result, judgmental assessment of the consequences of excitation—even relatively high excitation—can be very effective also. Such judgments are made primarily on the basis of field observations and laboratory research over a number of years and take into account the energy absorption characteristics (nonlinear behavior) inherent in many types of structural and mechanical systems.

In the design for seismic resistance, primary consideration is given to the main structural framing system(s). Consideration is given either explicitly or implicitly to the energy-absorbing effects of interior partitions, exterior covering, different types of materials, damping, drift, etc.

ATC-3 provisions were intended to be logically based with explicit consideration given to factors that are generally implicit in present code design provisions. Among the modern concepts presented in the ATC document that are significant departures from existing seismic building codes are the following:

1. Realistic seismic ground motion intensities.

2. Consideration of the effects of distant earthquakes on long period buildings.

3. Response modification coefficients (reduction factors) that are based on consideration of the inherent capacity for energy absorption, damping associated with inelastic response, and observed past performance of various types of framing systems.

4. The complexity of analysis and design dependent on importance or use factor, assigned building seismic performance category, and seismic motion intensity.

5. Simplified seismic response coefficient formulas related to fundamental period of building (one form of response spectra) but with specified restrictions.

6. Significant attention to the importance of connection details, structural redundancy and continuity.

7. Detailed requirements for architectural, electrical, and mechanical systems and components.

8. Material strengths at yield, i.e., a uniform limit stress design approach for all construction materials.

9. Guidelines for assessment and systematic abatement of seismic hazards in existing buildings.

10. Guidelines for assessment of earthquake damage, and strengthening or repair of damaged buildings and potential seismic hazards in existing buildings.

A summary of the most important aspects of the ATC-3 Tentative Provisions is contained in Tables 5 through 10. These tables are self-explanatory in general. The modal analysis procedure is not included in the tables since it is discussed in the text.

It should be noted that the provisions are intended to apply only to buildings and do not contain design requirements for special structures such as bridges, transmission towers, offshore structures, piers and wharves, industrial towers and equipment, and nuclear reactors.

TABLE 5. GUIDELINES PRINCIPLES

- Applicable to the entire United States

- Protect life safety

- Ensure functioning of essential facilities

- Allow for ingenuity of the designer

- Applicable to both new and existing buildings

- Include structural and nonstructural elements

- Applicable only to buildings

The ATC-3 design guidelines are centered around a set of principles that broadens, and thereby enhances, their usefulness to the design community, while at the same time these principles enable the designer to exercise ingenuity.

TABLE 6. ATC-3 CONCEPTS

- Realistic ground motion intensities

- Distant earthquake effects

- Response modification coefficients

- Analysis and design dependent on
    - Seismicity index
    - Importance or use
    - Building seismic performance

- Simplified building period calculation

- Nonstructural systems and components

- Strength design rather than working stress

- Seismic hazards in existing buildings

- Assessment of earthquake damage

- Response spectra

The ATC-3 provisions have incorporated modern dynamic analysis concepts. These tentative guidelines are presented explicitly, insofar as possible, in order to permit the structural designer to understand and evaluate the significance of the parameters employed in the design process.

TABLE 7. ATC-3 DESIGN STEPS

- Site location

- Map area number, A_a and A_v

- Seismicity index

- Seismic hazard exposure group and category

- Review of design requirements

    Soil profile
    Framing system
    Performance category
    Analysis procedure

- Analysis

    No design (wind governs)
    Equivalent lateral force (ELF) method
    Modal analysis method

Initial technical design steps are prescribed for all structures, regardless of the type of analysis procedure selected. ATC-3 guidelines provide two seismic risk maps for establishing the local probable (effective/design) acceleration, A_a, and the acceleration associated with velocity, A_v. The latter reflects possible effects of distant earthquakes on longer period structures.

TABLE 8. ELEMENTS OF EQUIVALENT LATERAL FORCE (ELF) ANALYSIS

- Base Shear

    $V = C_s W$

    where

    $$C_s = \frac{1.2 A_v S}{RT^{2/3}}$$ for R, see Table 9

    or

    $$C_s = \begin{cases} \dfrac{2.5 A_a}{R} \\ \\ \dfrac{2.0 A_a}{R} \end{cases}$$ when $A_a \geq 0.3$ for S_3 soil

    $T \leq 1.2 T_a$

    $$T_a = \begin{cases} C_T h_n^{3/4} \\ \\ \dfrac{0.05 h_n}{\sqrt{L}} \end{cases}$$

    for steel frame, $C_T = 0.035$
    for reinforced concrete frame, $C_T = 0.025$
    for all others

    $$S = \begin{cases} 1.0 \\ 1.2 \\ 1.5 \end{cases}$$
    for rock or stiff soil (S_1)
    for deep soil sites (S_2)
    for soft soils (S_3)

- Distribution of Shear Among Floors

    $$F_x = V \frac{w_x h_x^k}{\sum\limits_{i=1}^{n} w_i h_i^k}$$

    for $T \leq 0.5$, $k = 1$
    for $T \geq 2.5$, $k = 2$
    for $0.5 < T < 2.5$ interpolate

TABLE 8. (concluded)

- Distribution of Shear Within Each Floor
    Relative rigidity of resisting elements
    Torsional moments
    Accidental torsion

- Additional Considerations
    Overturning
    Drift
    $$\delta_x = C_d \delta_{xe} \qquad \text{for } C_d \text{ see Table 9}$$
    where
    $$\delta_{xe} = \text{Elastic deflection}$$

The ATC-3 equivalent lateral force analysis procedure incorporates response spectra coefficients and period expressions that are state-of-the-art. The procedures are not greatly different in form from those in other building codes already in existence.

TABLE 9. RESPONSE MODIFICATION COEFFICIENT R AND DEFLECTION COEFFICIENT C_d

Type of Structural System	Vertical Seismic Resisting System	R	C_d
1. Bearing Wall System (seismic resistance provided by shear walls or braced frames)	Light Walls, Shear Panels	6½	4
	Shear Walls		
	Reinforced Concrete	4½	4
	Reinforced Masonry	3½	3
	Braced Frames	4	3½
	Shear Walls		
	Unreinforced Masonry	1¼	1¼
2. Building Frame System (seismic resistance provided by shear walls or braced frames)	Light Walls, Shear Panels	7	4½
	Shear Walls		
	Reinforced Concrete	5½	5
	Reinforced Masonry	4½	4
	Braced Frames	5	4½
	Shear Walls		
	Unreinforced Masonry	1½	1½

TABLE 9. (concluded)

3. Moment Resisting Frame System (seismic resistance provided by ordinary or special moment frames)	Special Moment Frame Steel Reinforced Concrete Ordinary Moment Frame Steel Reinforced Concrete	8 7 4½ 2	5½ 6 4 2
4. Dual System (seismic resistance provided by special moment frames and shear wall or braced frames)	Shear Walls Reinforced Concrete Reinforced Masonry Wood Sheathed Shear Panels Braced Frames	8 6½ 8 6	6½ 5½ 5 5
5. Inverted Pendulum Structures (seismic resistance provided by isolated cantilever)	Special Moment Frame Structural Steel Reinforced Concrete Ordinary Moment Frame Structural Steel	2½ 2½ 1¼	2½ 2½ 1¼

The response modification coefficients and the deflection coefficients are used to establish seismic force levels and story drifts for different types of structural systems. The coefficients are an essential part of the ATC-3 design provisions and remain under evaluation as part of the ongoing trial design study program.

TABLE 10. SPECIAL DESIGN CONSIDERATIONS

- Redundancy

- Strength and stiffness discontinuities

- Story drift

- P-delta effects

- Review of design (and revision)

- Dynamic analysis

- Final design and details

- Quality assurance plan

The ATC-3 guidelines singled out a number of important design considerations for special attention.

Concluding Statement

Earthquake engineering is an evolving field of endeavor. The procedures and guidelines described herein for building analysis are believed to be representative of the best available at present, and assuredly will be improved in the years ahead as a result of research and practice. As in every aspect of engineering, the development of seismic design criteria and the use of analysis and design procedures and their resulting product, as well as construction practice and quality assurance, should be subjected to careful judgmental assessment at every stage. The goal is to produce safe and functional structures that can be constructed economically.

Acknowledgment

Much of the material contained in this monograph, and as presented in the EERI lecture series, was derived from the authors' research over the years, conducted in part under recent National Science Foundation Grants, from participation in the Applied Technology Council program, and from their personal studies associated with the development of seismic design criteria for nuclear power plants, for major pipeline projects, and for other types of industrial facilities. Any opinions, findings, conclusions or recommendations expressed herein are those of the authors and do not necessarily reflect the views of the National Science Foundation, the Earthquake Engineering Research Institute, the Applied Technology Council, the Nuclear Regulatory Commission, or any firm for which related studies have been conducted.

Figures 2, 3, 4, 7, 8, and 9 and portions of the accompanying text have appeared previously in Ref. 31. Significant portions of the last half of this monograph have been adapted from Chapter 2 (prepared by A. Chopra and N. M. Newmark) of Ref. 43. The authors acknowledge with gratitude the permission of the publishers to include the material noted.

Finally, the authors are indebted particularly to their colleagues Dr. A. K. Chopra and Dr. A. S Veletsos, whose contributions in various forms are reflected in the studies reported herein.

References

1. Schnabel, P. B. and H. B. Seed, "Acceleration in Rock for Earthquakes in the Western United States," *Bull. Seis. Soc. Amer.*, 63:2, April 1973, pp. 501-516.

2. Donovan, N. C., "A Statistical Evaluation of Strong Motion Data Including the February 9, 1971 San Fernando Earthquake," *Proc. 5th World Conf. on Earthquake Engineering*, Rome, 1974, Vol. 1, pp. 1252-1261.

3. Donovan, N. C. and A. E. Bornstein, "Uncertainties in Seismic Risk Procedures," *Jnl. Geotechnical Engineering Div.*, ASCE, 104:GT7, July 1978, pp. 869-887.

4. Campbell, K. W., "Near-Source Attenuation of Peak Horizontal Acceleration," *Bull. Seis. Soc. Amer.*, 71:6, Dec. 1981, pp. 2039-2070.

5. Idriss, I. M., "Characteristics of Earthquake Ground Motion," *Earthquake Engineering and Soil Dynamics*, ASCE, 1978, Vol. III, pp. 1151-1266.

6. Ambraseys, N. N., "Dynamics and Response of Foundation Materials in Epicentral Regions of Strong Earthquakes," *Proc. 5th World Conf. on Earthquake Engineering*, Rome, 1974, Vol. 1, pp. CXXVI-CXLVIII.

7. Trifunac, M. D. and A. G. Brady, "On the Correlation of Seismic Intensity Scales for the Peaks of Recorded Strong Ground Motion," *Bull. Seis. Soc. Amer.*, 65:1, Feb. 1975, pp. 139-162.

8. Newmark, N. M., W. J. Hall, and B. Mohraz, *A Study of Vertical and Horizontal Earthquake Spectra*, Directorate of Licensing, U.S. Atomic Energy Commission Report WASH-1255, April 1973, 151 p. (See also Newmark, N. M., J. A. Blume and K. K. Kapur, "Seismic Design Spectra for Nuclear Power Plants," *Jnl. Power Division*, ASCE, 99:P02, Nov 1973, pp. 287-303.)

9. Hall, W. J., B. Mohraz, and N. M. Newmark, *Statistical Studies of Vertical and Horizontal Earthquake Spectra*, Nuclear Regulatory Commission Report NUREG-0003, Jan 1976, 128 p.

10. Blume, J. A., R. L. Sharpe, and J. S. Dalal, *Recommendations For Shape of Earthquake Response Spectra*, Atomic Energy Commission Report WASH-1254, Feb 1973, 191 p.

11. Mohraz, B., *A Study of Earthquake Response Spectra for Different Geological Conditions*, Institute of Technology, Southern Methodist University, Dallas, Texas, 1975, 43 p.

12. Seed, H. B., C. Ugas, and J. Lysmer, *Site Dependent Spectra for Earthquake-Resistant Design*, EERC 74-12, Earthquake Engineering Research Center, University of California, Berkeley, 1974, 17 p.

13. Hanks, T. C., "Strong Ground Motion of the San Fernando, California Earthquake: Ground Displacements," *Bull. Seis. Soc. Amer.*, 65:1, Feb 1975, pp. 193-226.

14. Nuttli, O. W., "Similarities and Differences Between Western and Eastern United States Earthquakes, and Their Consequences for Earthquake Engineering," *Earthquakes and Earthquake Engineering—Eastern United States*, ed. J. E. Beavers, Ann Arbor Science, Vol. 1, 1981, pp. 25-51.

15. Idriss, I. M. et al, *Analyses for Soil-Structure Interaction Effects for Nuclear Power Plants*, Committee on Nuclear Structures and Materials of the Structural Division, ASCE, 1979, 155 p.

16. Johnson, J. J., *Soil-Structure Interaction: The Status of Current Analysis Methods and Research*, Seismic Safety Margins Research Program, Lawrence Livermore Laboratory Report for the U.S. Nuclear Regulatory Commission, NUREG/CR-1780, 1981.

17. Cloud, W. K., "Intensity Map and Structural Damage, Parkfield, California Earthquake of June 27, 1966," *Bull. Seis. Soc. Amer.*, 57:6, Dec 1967, pp. 1161-1178.

18. *Managua, Nicaragua Earthquake of December 23, 1972*, Earthquake Engineering Research Institute Reconnaissance Report, May 1973, 214 p. (See also 2 Vols. Conf. Proceedings by EERI.)

19. *The San Fernando, California Earthquake of February 9, 1971*, U. S. G. S. Professional Paper 733, 1971, 254 p.

20. *Reconnaissance Reports—Imperial Valley Earthquake, October 15, 1979*, NRC Reconnaissance Team, Nov 1979.

21. *Reconnaissance Report, Mikagi—Ken—Oki, Japan Earthquake, June 12, 1978*, Earthquake Engineering Research Institute, Dec 1978, 165 p.

22. *Engineering Features of the San Fernando Earthquake, February 9, 1971*, ed. P. C. Jennings, California Institute of Technology, Earthquake Engineering Research Laboratory, EERL 71-02, June 1971, 512 p.

23. "Seismological Notes," *Bull. Seis. Soc. Amer.*, 63:1, Feb 1973, pp. 335-338.

24. "Seismological Notes," *Bull. Seis. Soc. Amer.*, 63:3, June 1973, p. 1177-1183.

25. *Proc. 2nd U.S. National Conf. on Earthquake Engineering*, Earthquake Engineering Research Institute, 1979.

26. Kennedy, R. P., "Peak Acceleration As a Measure of Damage," *Proc. 4th International Seminar on Extreme-Load Design of Nuclear Power Facilities*, Paris, France, Aug 1981.

27. Newmark, N. M. and E. Rosenblueth, *Fundamentals of Earthquake Engineering*, Prentice-Hall, Englewood Cliffs, N. J., 1971.

28. Clough, R. W. and J. Penzien, *Dynamics of Structures*, McGraw-Hill, New York, 1975.

29. Blume, J. A., N. M. Newmark, and L. Corning, *Design of Multi-Story Reinforced Concrete Buildings for Earthquake Motions*, Portland Cement Assoc., Chicago, 1961, 318 p.

30. Newmark, N. M. and W. H. Hall, *Procedures and Criteria for Earthquake Resistant Design*, Building Practices for Disaster Mitigation, National Bureau of Standards, Washington, D.C., Building Sciences Series 46, Vol. 1, Feb 1973, pp. 209–236.

31. Newmark, N. M., "Earthquake Response Analysis of Reactor Structures," *Nuclear Engineering and Design* (The Netherlands), 20:2, July 1972, pp. 303–322.

32. Newmark, N. M. and W. J. Hall, *Development of Criteria for Seismic Review of Selected Nuclear Power Plants*, Nuclear Regulatory Commission Report NUREG/CR-0098, May 1978, 49p.

33. Newmark, N. M. and W. J. Hall, "Vibration of Structures Induced by Ground Motion," in *Shock and Vibration Handbook*, eds. C. M. Harris and C. E. Crede, McGraw-Hill, New York, 2nd Ed., 1976, pp. 29-1 to 29-19.

34. Newmark, N. M., "Current Trends in the Seismic Analysis and Design of High Rise Structures," Chapter 16 in *Earthquake Engineering*, ed. R. L. Wiegel, Prentice-Hall, Englewood Cliffs, N. J., 1970, pp. 403–424.

35. Housner, G. W., "Design Spectrum," Chapter 5 in *Earthquake Engineering*, ed. R. L. Weigel, Prentice-Hall, Englewood Cliffs, N. J., 1970, pp. 93–106.

36. Jacobsen, L. S. and R. S. Ayre, *Engineering Vibrations*, McGraw-Hill, New York, 1958.

37. Veletsos, A. S. and N. M. Newmark, "Response Spectra of Single-Degree-Of-Freedom Elastic and Inelastic Systems," in *Design Procedures for Shock Isolation Systems of Underground Protective Structures*, Report RTD TDR-63-3096, Vol. III, Air Force Weapons Laboratory, Kirtland Air Force Base, New Mexico, June 1964, 220 p.

38. Veletsos, A. S., N. M. Newmark, and C. V. Chelapati, "Deformation Spectra for Elastic and Elastoplastic Systems Subjected to Ground Shock and Earthquake Motions," *Proc. 3rd World Conf. on Earthquake Engineering*, New Zealand, 1965, Vol. II, pp. 663–682.

39. Nigam, N. C. and P. C. Jennings, "Calculation of Response Spectra From Strong-Motion Earthquake Records," *Bull. Seis. Soc. Amer.*, 59:2, April 1969, pp. 909-922.

40. Newmark, N. M. and R. Riddell, "Inelastic Spectra for Seismic Design," *Proc. 7th World Conf. on Earthquake Engineering*, Istanbul, Turkey, 1980, Vol. 4, pp. 129-136.

41. *Tentative Provisions for the Development of Seismic Regulations for Buildings*, Applied Technology Council ATC-3-06, Nat. Bur. of Stds. Spec. Publ. 510, 1978.

42. Chopra, A. K., *Dynamics of Structure—A Primer*, Earthquake Engineering Research Institute, Berkeley, California, 1981.

43. Chopra, A. K. and N. M. Newmark, "Analysis," Chapter 2 in *Design of Earthquake Resistant Structure*, ed. E. Rosenblueth, Pentech Press Ltd., London, 1980.

44. *Recommended Lateral Force Requirements and Commentary*, Seismology Committee, Structural Engineers Association of California, 1968 edition.

45. *Recommended Lateral Force Requirements and Commentary*, Seismology Committee, Structural Engineers Association of California, 1975 edition.